Fundamental Collagen Chemistry in Leather Making
制革基础胶原化学

林 炜　石佳博　著
〔德〕Günter Reich

四川大学出版社

项目策划：王　锋
责任编辑：王　锋
责任校对：周维彬
封面设计：墨创文化
责任印制：王　炜

图书在版编目（CIP）数据

制革基础胶原化学 = Fundamental Collagen Chemistry in Leather Making：英文 / 林炜，石佳博，（德）金特·赖希（Günter Reich）著．— 成都：四川大学出版社，2020.8
ISBN 978-7-5690-3465-3

Ⅰ．①制… Ⅱ．①林… ②石… ③金… Ⅲ．①制革化学－胶原蛋白－高等学校－教材－英文　Ⅳ．① TS513

中国版本图书馆 CIP 数据核字（2020）第 158773 号

书名	制革基础胶原化学
	ZHIGE JICHU JIAOYUAN HUAXUE
著　者	林　炜　石佳博　〔德〕Günter Reich
出　版	四川大学出版社
地　址	成都市一环路南一段 24 号（610065）
发　行	四川大学出版社
书　号	ISBN 978-7-5690-3465-3
印前制作	四川胜翔数码印务设计有限公司
印　刷	四川盛图彩色印刷有限公司
成品尺寸	185mm×260mm
插　页	4
印　张	12
字　数	573 千字
版　次	2021 年 1 月第 1 版
印　次	2021 年 1 月第 1 次印刷
定　价	58.00 元

版权所有　◆　侵权必究

◆ 读者邮购本书，请与本社发行科联系。
　电话：(028)85408408/(028)85401670/
　(028)86408023　邮政编码：610065
◆ 本社图书如有印装质量问题，请寄回出版社调换。
◆ 网址：http://press.scu.edu.cn

四川大学出版社
微信公众号

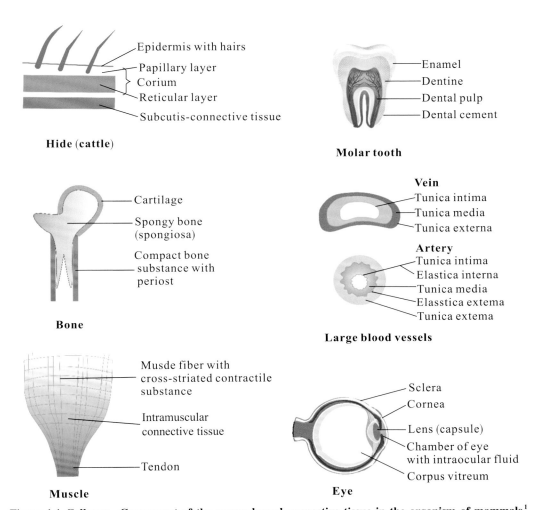

Figure 1.1 Collagen: Component of the mesenchymal connective tissue in the organism of mammals [1]

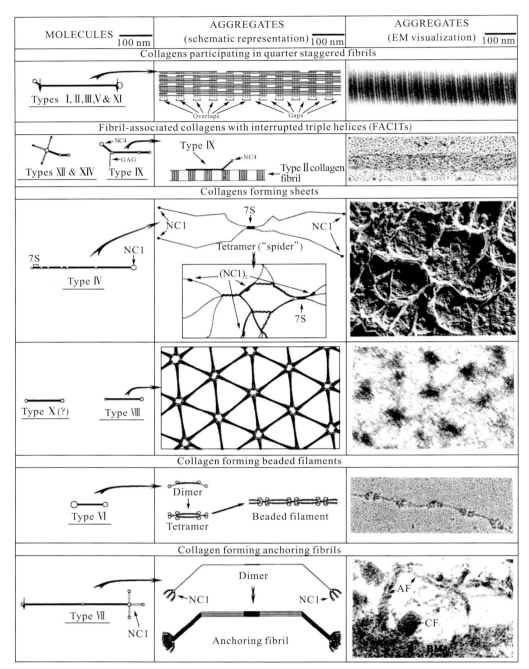

Figure 1.2 The collagen family. Reprinted with permission from *The FASEB Journal*, 1991, 5(13): 2814-2823.

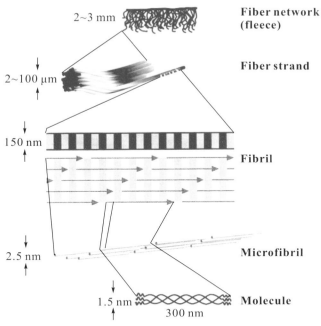

Figure 1.3 Structural hierarchy of Type I collagen

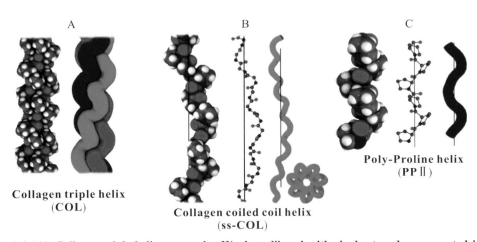

Figure 1.4 (A) Collagen triple helix as van der Waals radii and with single strands represented in different colors. (B) Single polypeptide chain as it is within the collagen helix, ball and stick (backbone atoms only), and tube (side and top views) representations. (C) Single polypeptide chain (all proline) in the PP II conformation, ball and stick, and tube representation. Color code: carbon/cyan, oxygen/red, nitrogen/blue, and hydrogen/light gray. Reprinted with permission from *The Journal of Physical Chemistry B*, 2019, 123(34): 7354-7364. Copyright (2019) American Chemical Society.

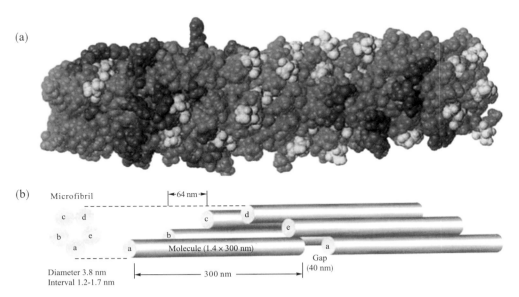

Figure 1.6 (a) Computer model and (b) structural principle and molecular arrangement of the Smith microfibril. Reprinted with permission from *Journal of the American Leather Chemists Association*, **1997**, 92(8): 185-198.

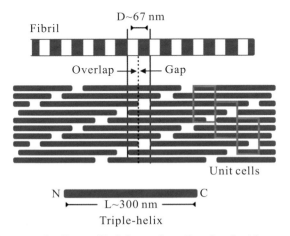

Figure 1.7 Schematic diagram of collagen fibril formation. Reprinted with permission from *Biophysical Journal*, **2016**, 111(1): 50-56. Copyright (2012) Elsevier.

Figure 1.10 Collagen fibers in three-dimensional weave (polarized light image of a pigskin section)

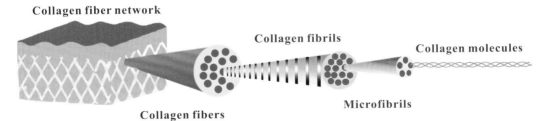

Figure 2.3 Principle of the structural composition of Type I collagen

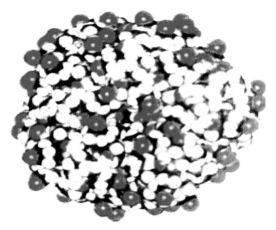

Figure 2.4 Cross-section of the computer model of a microfibril

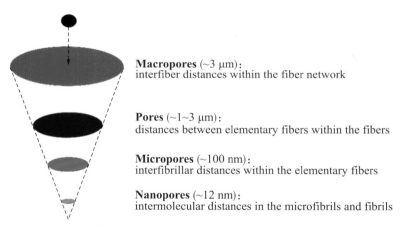

Macropores (~3 μm):
interfiber distances within the fiber network

Pores (~1~3 μm):
distances between elementary fibers within the fibers

Micropores (~100 nm):
interfibrillar distances within the elementary fibers

Nanopores (~12 nm):
intermolecular distances in the microfibrils and fibrils

Figure 2.6 "Diffusion funnel" of collagen

Preface

Leather is a material manufactured from the collagen fiber network of hides and skins, provided with the characteristics required for the intended purpose, and produced by suitable chemical and physical processes. As this definition clearly shows, collagen constitutes the raw material for the manufacture of leather.

Collagen is the most common regenerative biopolymer, along with cellulose and lignin. Animal hides or skins contain dominantly type I collagen and are commercially important as natural fiber network largely utilized in traditional leather industry. As an important constituent of the mesenchyme and the extracellular matrix, collagen plays a significant role in healthy living organisms and pathological conditions. Therefore, collagen has been one of the most widely investigated proteins nowadays, not only drawing attention from food and leather industries, but also current emerging biomedical applications. This is illustrated by the fact that more than 30 different types of collagen and recombinant collagen have now been identified.

Type I collagen plays a key role in the manufacture of leather and in determining the properties of leather. Hence there is a tremendous need to make use of the enormous increase in our knowledge about the structure and reactivity of collagen in order to gain an even greater understanding of the processes that take place when skin is converted into leather, which is essential for the development of leather manufacturing technology and new leather auxiliaries now and future. This is the purpose of our studies that have been performed in past years. At the same time, we intend to show what changes the collagen is subjected to during its transformation into leather, and how the latter properties are influenced by the collagen.

Some results of these studies have now been collected in this book in order to make them available to a wider public. The respective chapters are self-contained and can be read independently of each other, depending on the reader's interests and the time available. All of these studies essentially focus on processes that involve collagen.

Chapter 1 provides a general introduction and answers the question "What is leather?" which is an issue that is also of interest to those who are not directly concerned with leather. Chapter 2 covers the diffusion and binding processes in which collagen is involved and takes a quantitative approach that has rarely been adopted in the past. The authors are aware that many questions still have to be answered in order for progress to be made, but they believe that this methodology will enable new approaches to be taken in future. Chapter 3 contains a critical appraisal of the theory of tanning from its beginnings through to the present day and speculates on future developments. Many questions still need to be answered here, even though this may involve challenging opinions that have long been established in the textbooks. Chapter 4 emphasizes that water and collagen form an integrated system; and water plays a key role in the manufacture of leather and determines many different properties of leather, which may be overlooked often. Chapter 5 summarizes the fundamental nature of various typical tanning, including vegetable, mineral, oil, aldehyde tanning, as well as syntans and combination tanning from the perspectives of tanning agents and tanning mechanisms.

In addition, the Appendix includes 3 parts: common leather vocabularies excerpted from current International Standard "*ISO/FDIS 15115 Leather-Vocabulary*", British Standard "*EN15987 Leather-Terminology-Key definitions for the leather trade*" and "*BS2780:1983 +A1: 2013 Glossary of leather terms*"; the word-forming patterns for the vocabularies related to leather-making and leather chemicals; and our representative research articles concerning tanning reactions and technologies published in recent decade.

This book is proved to be appropriate for being the textbook for 2nd and 3rd year undergraduates since the main paragraphs has been introduced to students for 10 years. It is also a reference material for postgraduate students to review and relearn collagen chemistry in leather-making. In fact, the existing modification methods applied in collagen-based materials, all without exception is in accordance with the fundamental tanning chemistry. It is to be hoped that readers would be stimulated by the enthusiasm of the authors for leather and the collagen from which it is made. Also, it would be greatly appreciated to readers for the suggestions and criticism so that this book can be improved in the future. We thank teachers, colleagues and students who reference and adopt this text book.

Prof. Dr. Günter Reich has been involved with the theory and practice of collagen all his working life, and he is very grateful to the colleagues from BASF AG for working together with them happily. And Prof. Dr. Wei Lin wishes to use this chance to sincerely thank the Department of Biomass and Leather Engineering, Sichuan University, and her colleagues and students for their support, understanding, encouragements and helps in years.

Authors

December 26, 2020

Contents

Chapter 1　The Structure and Reactivity of Collagen ……………………………… (1)
　1.1　Occurrence and Industrial Significance of Collagen ……………………… (1)
　1.2　What is Collagen ……………………………………………………………… (3)
　1.3　How is Collagen Formed …………………………………………………… (16)
　1.4　The Role of Water in Collagen ……………………………………………… (17)
　1.5　The Reactivity of Collagen …………………………………………………… (18)
　1.6　The Nature of Leather Formation and the Characteristics of Leather ……… (19)
　1.7　Concluding Remarks ………………………………………………………… (23)

Chapter 2　Physical and Chemical Processes on Collagen and Its Transformation into the Leather Matrix …………………………………………… (24)
　2.1　Introduction …………………………………………………………………… (24)
　2.2　Preliminary Remarks ………………………………………………………… (26)
　2.3　Principles of Tanning Theory, Basic Assumptions, Definitions and Methods …… (26)
　2.4　Diffusion and Penetration …………………………………………………… (31)
　2.5　Material Binding in and on the Collagen …………………………………… (34)
　2.6　Interaction Equivalents of Active Substances with Collagen ……………… (36)
　2.7　Summary ……………………………………………………………………… (48)

Chapter 3　The Theory of Tanning—Past, Present, Future ……………………… (49)
　3.1　Introduction …………………………………………………………………… (49)
　3.2　The History of Tanning Theory ……………………………………………… (51)
　3.3　Current Knowledge of the Nature of Tanning ……………………………… (60)
　3.4　Tanning Theory in the Future ……………………………………………… (71)

Chapter 4　The Significance of Water for the Structure and Properties of Collagen and Leather ……………………………………………………………… (74)
　4.1　Introduction …………………………………………………………………… (74)
　4.2　Collagen and Water: An Integral System …………………………………… (75)
　4.3　The Consequences for the Collagen Structure of a Change in Water Content …… (76)

Fundamental Collagen Chemistry in Leather Making

 4.4 Water Determines the Collagen Characteristics ……………………………… (77)
 4.5 Thermal and Hydrothermal Stability of Collagen and Leather ……………… (77)
 4.6 The (Hydro)Thermal Stability of Collagen and Leather as a Function of the
   Water Content and Tanning ……………………………………………………… (79)
 4.7 The Role of Water in the Conversion of Collagen to Leather ………………… (81)
 4.8 Tanning as Dehydration: A Chapter of Tanning Theory ……………………… (82)

Chapter 5 Introduction of Modern Tanning Chemistry ……………………… (86)
 5.1 Vegetable Tanning ………………………………………………………………… (86)
 5.2 Mineral Tanning …………………………………………………………………… (89)
 5.3 Oil Tanning ………………………………………………………………………… (95)
 5.4 Aldehyde Tanning ………………………………………………………………… (96)
 5.5 Syntans ……………………………………………………………………………… (100)
 5.6 Combination Tanning ……………………………………………………………… (101)

Biography ……………………………………………………………………………………… (105)

Appendix Ⅰ …………………………………………………………………………………… (116)
 Process ………………………………………………………………………………… (116)
 Material ……………………………………………………………………………… (117)
 Equipment …………………………………………………………………………… (120)
 Performance ………………………………………………………………………… (121)

Appendix Ⅱ …………………………………………………………………………………… (122)
 Prefix ………………………………………………………………………………… (122)
 Suffix ………………………………………………………………………………… (127)

Appendix Ⅲ …………………………………………………………………………………… (136)
 Representative Research Articles ………………………………………………… (136)

Chapter 1   The Structure and Reactivity of Collagen

The many manufacturing processes, types and properties of leather make definition of the material difficult. One possible definition is as follows[1]:

"Leather is a material manufactured from the collagen fiber network of hides and skins, provided with the characteristics required for the intended purpose, and produced by suitable chemical and physical processes."

As this definition clearly shows, collagen constitutes the raw material for the manufacture of leather[2]. A description of its structure and reactivity is the purpose of the present study. At the same time, however, we intend to show what changes the collagen is subjected to during its transformation into leather, and how the latter's properties are influenced by the collagen. Following this excursus, the definition of "leather" can be formulated somewhat more precisely in terms of the relationship between its structure and properties.

1.1   Occurrence and Industrial Significance of Collagen

Collagen (the term derives from the Greek and means "glue-forming") is by a comfortable margin the most widely found protein in terms of quantity[3], and together with cellulose and lignin, one of the three quantitatively dominant biopolymers[4]. This fact is due to its incidence: It is present throughout the entire animal kingdom with the exception of single-cell organisms. Its significance derives from its functions in living organisms. Mammals are a good example: As a component of the mesenchymal connective tissue, collagen is a primary component of the skeleton and the skin, and thus responsible for protecting and supporting the body[5-7]; it is, however, represented in numerous other organs, and as part of the extracellular matrix (ECM)[8-10], is of great significance for cellular biology (Figure 1.1).

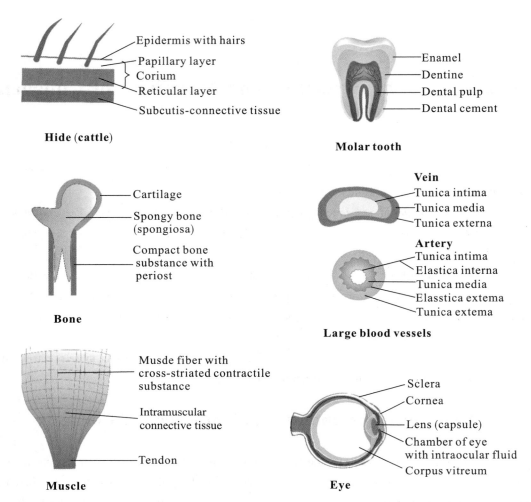

Figure 1.1 Collagen: Component of the mesenchymal connective tissue in the organism of mammals[1]

Once mocked in biochemical research as "boring" and being of interest at most to leather chemists for their own purposes, collagen now attracts wide attention in many scientific disciplines, and is in fact one of the most closely studied proteins[11-13]. This can be attributed to the fact that the role played by collagen in cellular processes, in the growth of connective tissue[14-16], and also in the incidence of certain clinical conditions ("collagen diseases") and for the natural ageing of the organism (a human being is as old as his or her connective tissue), has increasingly been recognized.

Collagen is also of economic significance. Hides and skins, primarily those of cattle, sheep, goats and pigs, are by-products of human foodstuff and wool production. Collagen is thus a renewable raw material the production of which grows in proportion to the cattle stock, which in turn is growing in proportion to the earth's human population. Despite certain changes in eating habits, the trend towards vegetarian diets and white meats (poultry and fish), the Bovine Spongiform Encephalopathy (BSE) crisis, and the shift in population growth to regions of low meat consumption, the ratio of cattle stock to world human population has proved relatively stable.

At approximately 4000 kt per annum, consumption of collagen exceeds that of wool at 1520 kt per annum, substantially exceeds that of casein at around 100 kt per annum, and in fact equates to half of the world production of polyurethanes. It encompasses leather, by a wide margin the most significant area of consumption, gelatines and glue, casings (collagen-based sausage casings/collagen films), hydrolysates, carcass meal and fertilizers, and special industrial medical products (employed under the heading of "tissue engineering"[17]). In the last of these cases, the production of recombinant collagen is increasing in importance (Table 1.1).

Table 1.1 Annual production of materials from collagen[1]

Collagen product		Quantity/Proportion
Light leather		1500 million m^2
	Comprising cattle	925 million m^2
	Comprising sheep/goats	380 million m^2
	Comprising pigs	200 million m^2
Heavy leather		450 kt
Lighter leather (total)		100%
	Comprising shoes	60%
	Comprising garments	15%
	Comprising furniture and automotive	12%
	Comprising fancy goods	13%
Leather board		100 kt
Gelatines/glue		300 kt
Casings (sausage casings etc.)		>1 billion m
Hydrolysates		200 kt
Carcass meal (animal feed/fertilizer)		>2 million tons
Collagen in human foodstuffs		800 kt

1.2 What is Collagen

Collagen is now known to be not a single substance but rather a family of substances[18-20]. The collagen types now known, almost 30 in number, differ in some cases considerably in their chemical and structural characteristics according to their function in the living organism[21]. A common feature is their conformance (at least partial) to a right-handed triple-helical structure ("triple-helix domains"), their hydroxyproline content, an amino acid which is virtually specific to collagen and their occurrence in the extracellular matrix. Collagens include the fibril-forming types, types which control fibril growth and are associated with the fibrils, and also structures which form networks and filaments[24]. The relationships between the structure of these different collagens and their significance within the living organism are now well documented (Figure 1.2)[25].

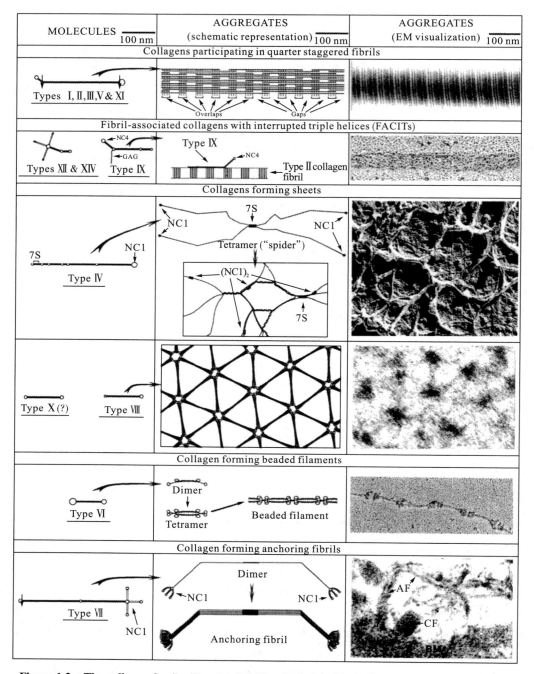

Figure 1.2 The collagen family. Reprinted with permission from *The FASEB Journal*, 1991, 5 (13): 2814-2823.

The corium of animal hide, which is processed in leather manufacture, also contains several of these collagens. Fibril-forming Type Ⅰ collagen dominates by a wide margin as the most abundant collagen type. All further discussion will thus relate to this type (Table 1.2).

Table 1.2 Collagen types in animal hide[1]

Type	Properties	Remarks
Ⅰ	Fibril-forming, growth extending to fibers and fiber network	Principal components (95%), crucial to the properties
Ⅲ, Ⅴ	Forms fine fibrils, fibers and fiber strands	Enriched in the grain layer (in particular Type Ⅲ), in some cases microfibers with Type Ⅰ; overall, not crucial to the properties
Ⅳ, Ⅶ	Not fibrillar (Type Ⅳ), filamentous	In basement membrane, interface between epidermis and corium, "enzyme-sensitive" (enzymatic dehairing)
Ⅵ	Filamentous	Regulatory for fibril assembly, removed in the liming

1.2.1 Amino Acid Composition and Sequence of Type Ⅰ Collagen

Like keratin, elastin and silk fibroin, Type Ⅰ collagen is a scleroprotein, consisting like all proteins of amino acids linked by —CO—NH— groups to form peptide chains. Of the three chains forming collagen, two are identical in the case of Type Ⅰ collagen, α1 (Ⅰ), and one differs in its amino acid composition: $[\alpha1(Ⅰ)]_2(\alpha2)$. The α chains which combine to form the triple helix exhibit non-helical and short-chain appended peptides (telopeptides) at their N- and C-terminal ends[26, 27]. These may be disregarded for the purpose of the present discussion. A chain consists of 1014 amino acids and includes, together with two Hyp isomers, 19 different amino acids, of which the two acidic amino acids (Glu and Asp) are present in part in the form of amides (Gln and Asn). For the sake of illustration, the number of amino acids per molecule was calculated from data on the percentage content of individual amino acids in the (α1) chain (number of amino acids per 100 amino acids). A total number of 3 × 1014 = 3042 amino acids per molecule was assumed for this purpose. The amino acids with the different functional groups (acidic, basic, hydroxyl groups and peptide groups) are marked in color. The term "functionality" encompasses ionic, covalent, complex and hydrogen bonding.

Of the 19 different amino acids involved in the creation of collagen, a third is accounted for by glycine (Gly)[28], and almost a further third by the cyclic amino acids Proline (Pro) and Hydroxyproline (Hyp). This is a requirement for the collagen-specific formation of a triple helix. As already mentioned, Hyp is virtually specific to the structural stability collagen[29-31], a fact which can be exploited for its quantitative detection. It is also substantially responsible for the hydrothermal stability of native, fully hydrated collagen, the latter undergoes a helix-coil transition at defined temperatures (shrinkage temperature, noted as T_s), i. e. it denatures[32]. The functionality is due partly to the content of acidic amino acids Glu and Asp. Should these originally be available as amides (Gln and Asn), they undergo partial deamination (Gln to Glu, Asn to Asp) in the liming process to approximately 25%~50%. The functionality is further due to the content of the basic amino acids Lys, Hyl, Arg and His and of the hydroxylated amino acids Hyp, Hyl, Ser and Tyr. In the past, the latter have often been considered less relevant to tanning; this view should however be reconsidered. The peptide groups in the fundamental chain are also important for the reactivity of the collagen. If present as a fiber network, as in the case of hide (and leather), denaturing is manifested by shrinkage to a third of the original length. The T_s at which this transformation takes place constitutes a measure of the hydrothermal stability. For example, T_s of fish skin collagen is dependent upon its hydroxyproline content, and that the T_s corresponds to the temperature of seawater in which the fish concerned live.

The functionally equivalent amino acids were added and the corresponding bond equivalents calculated in milliequivalents (m_{Eq}) per gramme of collagen based upon a relative molar mass of Type I collagen of 300 kD, corresponding to a mean relative molar mass of 98.6 per amino acid:

m_{Eq}/g collagen = number of amino acids in the molecule: 300000 × 1000

For the acidic amino acids, a maximum value (complete deamidation: ΣAsp, Asn, Glu, Gln), a minimum value according to the amino acid composition (ΣAsp, Glu) and a value relevant to tanning chemistry are indicated. The last of these values assumes a (mean) deamidation of 25%.

The content of acidic and basic amino acids, amino acids containing OH groups and peptide groups is important, and is decisive for the reactivity of the collagen, which is a requirement for its transformation into leather (Table 1.3). If stoichiometric aspects are to be added to this functionality, the accessibility of the reactive groups must be taken into account. The peptide groups are involved substantially in the stabilization of the triple helix and are not therefore available for hydrogen bonding which is essential for the triple helical conformation of the collagen, with other substances. Attention is also drawn to the not inconsiderable amidation of the carboxylic acids.

Chapter 1 The Structure and Reactivity of Collagen

Table 1.3 Type I collagen: Content of amino acid crucial to the structure and reactivity[1]

Amino acids	Content in %	Number per molecule	Types
Gly	33.4	1016	
Pro	12.9	392	
Ala	10.5	319	
Hyp	9.2	280	Hydroxyl function
Glu	4.6	140	Acidic function
Gln	2.6	79	Acidic function
Arg	4.8	146	Basic function
Asp	3.5	106	Acidic function
Asn	1.3	40	Acidic function
Ser	3.8	116	Hydroxyl function
Leu	2.5	76	
Lys	2.5	76	Basic function
Val	1.9	58	
Thr	1.7	52	
Phe	1.3	40	
Ile	1.1	33	
Hyl	0.7	21	Both basic and hydroxyl function
Met	0.7	21	
His	0.5	15	Basic function
Tyr	0.5	15	Hydroxyl function
—CO—NH—	99	3039	

Buttar *et al.* provide data here for the acidic, basic and hydroxyl groups[33]; Heidemann discusses the peptide groups. The latter assumes that only one third of the available peptide groups are accessible, as the remainder are involved in stabilization of the hydrogen bonding of the triple helix and are not therefore available. He also points out that owing to the short distance of the peptide groups of 0.32 nm along the helix, a certain number are always unavailable as "functional" peptide groups for steric reasons when the functional groups in corresponding co-reactants (e. g. the phenolic hydroxyl groups in vegetable and synthetic tanning agents) possess an interval exceeding 0.32 nm. Heidemann postulates from this a further reduction in the reactive peptide groups by a factor of 10 (Table 1.4).

7

Fundamental Collagen Chemistry in Leather Making

Table 1.4 Availability of functional groups based upon Buttar et al. and Heidemann

Dimension	Acidic groups	Basic groups	Hydroxyl groups	Peptide groups
Present (in %)	75	100	100	100
Available following partial deamidation (in %)	100			
Actually accessible (in %)	75	77	61	
Available minus its own hydrogen bond formation (in %)				33
Still available following steric obstruction (in %)				4

The restricted accessibility of functional groups for the peptide groups as inferred by Buttar *et al.* from the computer model of the microfibril and postulated by Heidemann raises questions regarding the situation in and on the fibrils. It can be assumed that the molecular arrangement there is subject to the same conditions and distances as in the (more or less hypothetical) microfibril. This would however have far-reaching consequences regarding the chemical bond equivalents of the collagen in general, and gives rise to contradictions with the experimental findings, for example the acidic and basic bond equivalents derived from titration curves.

Attention is also drawn to the known pH dependency of the availability of reactive groups according to their respective dissociation. In addition, the different pK values of Asp and Glu have an influence upon complex formation for example with chromium salts. Consideration of this area has been postponed for the time being.

Leaving aside this need for clarification, the bond equivalents derive from the above explanations are summarized below. The chemical bond equivalents of the collagen thus determined which are actually employed for calculation of the bonding capacities of Type I collagen for the most important classes of substances are shown in Table 1.5.

Table 1.5 Number of functional groups available or accessible in Type I collagen, per mol, and resulting bond equivalents in m_{Eq}/g collagen

Group type	Associated amino acid or group	Number /molecule	m_{Eq}/g collagen
Acidic and amidated acidic amino acids (—COOH, —CONH$_2$) (theoretical maximum value)	Gln, Glu, Asn, Asp	365	1.15
Acidic (—COOH) amino acid (theoretical minimum value)	Glu, Asp	246	0.78
Acidic (—COOH) amino acid (following 25% deamination, value relevant to tanning chemistry)	Glu, Asp	308	1.03
According to surface area availability (value from model)	Glu, Asp	231	0.77
Basic (—NH$_2$) amino acid (value relevant to tanning chemistry)	Lys, Hyl, Arg, His	258	0.86

Group type	Associated amino acid or group	Number /molecule	m_{Eq}/g collagen
According to surface area availability (value from model)	Lys, Hyl, Arg, His	199	0.66
Hydroxyl-functional amino acids (value relevant to tanning chemistry)	Hyp, Hyl, Ser, Tyr	432	1.44
According to surface area availability (value from model)	Hyp, Hyl, Ser, Tyr	263	0.88
—CO—NH—	All peptide groups	3039	10.13
—CO—NH—	Minus groups following steric obstruction	1033	3.34
—CO—NH—	Minus groups which have formed hydrogen bonds themselves	122	0.40
Sum of all potential hydrogen bond-forming agents	All petide groups, all basic and hydroxyl functional amino acids	3729	12.43

As far as the sequence of the approximately 1050 amino acids with peptide-like linkage per α chain are concerned, two aspects must be considered which are important for the structure and the reactivity of collagen. Firstly, Gly must always be located in the third position. The α chains must therefore be regarded as a regular recurrence of the sequence $(Gly-X-Y)_n$, where n is approximately 350 according to the chain length of Type I collagen α = $(Gly-X-Y)_{350}$ in which X is frequently proline and hydroxyproline is always located in position Y. The tripeptide sequence (Gly-X-Y) is characteristic for the amino acid sequence of Type I collagen, Pro frequently representing X and Hyp Y. At 11.80% (> 11.8), tripeptides of the type (Gly-Pro-Hyp) are not uncommon; (Gly-Pro-Ala) and (Gly-Ala-Hyp) also occur relatively frequently, at 8.9% and 6.2% respectively. The amino acid sequence (Gly-X-Hyp) is substantially responsible for the specific helix structure of the collagen[34,35]. Synthetically manufactured polypeptides with this structure therefore represent preferred model substrates for structure and reactivity studies of Type I collagen.

The second aspect concerns the fact, of importance to the reactivity of the collagen and therefore also to fixation of the tanning material, that the reactive acidic and basic amino acids in the α chains are distributed by no means statistically according to their number, but occur in clusters. They impart a certain charge pattern to the collagen, which is a requirement for the assembly of fibrils. The α chains of the collagen arrange to form higher structures, giving rise to a proper structural hierarchy of the collagen (Figure 1.3). Three α chains thus initially combine to form a three-stranded "triple helix" with a rise of 0.29 nm per amino acid residue and approximately 3.3 amino acid residues per turn. The typical feature of collagen is an elegant structural motif in which three parallel polypeptide strands in a left-handed, polyproline II -type

(PP II) helical conformation coil about each other with a one-residue stagger to form a right-handed triple helix (Figure 1.4) [36, 37]. Owing to its solubility, the collagen can be characterized in solution by scattered light measurement and by means of ultracentrifuging[38]. Isolated from solution, it can be viewed and measured by means of transmission electron microscopy (TEM) and atomic force microscopy (AFM) [39-41]. A collagen molecule ("tropocollagen") soluble in acid and/or citrate buffer is thus formed the dimensions of which are now familiar and which can be viewed by means of AFM (Figure 1.5). The collagen molecule is 300 nm in length and has a diameter of 14 nm[42]. It consists of three α chains which are helically intertwined and which form the triple helix of the collagen molecule. The triple helix is stabilized by intramolecular hydrogen bonds between the amino acids, and also by the involvement of water. For steric reasons, the presence of Gly at each third position is a requirement for the triple-helix structure, and gives rise to the close proximity of the α chains within the helix[43]. Space is therefore available within the helix structure for molecules with the dimensions of the water. Their intramolecular penetration would thus require displacement of the water, resulting in destabilization of the triple helix. The intramolecular distances are in the order of 0.15 nm.

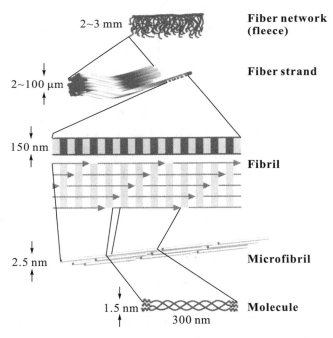

Figure 1.3 Structural hierarchy of Type I collagen[1]

Chapter 1　The Structure and Reactivity of Collagen

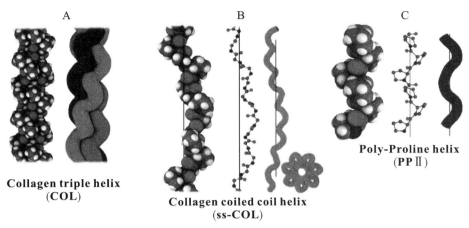

Figure 1.4　(A) Collagen triple helix as van der Waals radii and with single strands represented in different colors. (B) Single polypeptide chain as it is within the collagen helix, ball and stick (backbone atoms only), and tube (side and top views) representations. (C) Single polypeptide chain (all proline) in the PP Ⅱ conformation, ball and stick, and tube representation. Color code: carbon/cyan, oxygen/red, nitrogen/blue, and hydrogen/light gray. Reprinted with permission from *The Journal of Physical Chemistry B*, 2019, 123(34): 7354-7364. Copyright (2019) American Chemical Society.

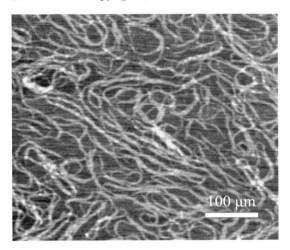

Figure 1.5　AFM image of the collagen molecule

1.2.2　The Microfibril

Fibrils are assembled from collagen molecules in vivo/vitro (the process being triggered in the latter case by a change in pH and/or a rise in temperature). The Smith microfibrils constitute a metastable intermediate stage. The microfibrils in this case are successive layers of five molecules forming a "pentagon". Similar aggregates were postulated in the past comprising three and more recently eight molecules[44]. As the Smith microfibril has frequently been the subject of computer modelling of the fixation of chrome tanning material and is only an intermediate stage[45,46], it is not crucial to an understanding of fibril formation whether it is a

Fundamental Collagen Chemistry in Leather Making

pentagon or an octagon, and the Smith model will be assumed here (Figure 1.6)[47]. Tuckerman was the first to demonstrate, by means of AFM during in vitro assembly experiments, the actual existence of the microfibrils as an intermediate stage in fibril formation. Although its composition is not undisputed, the Smith microfibril constitutes only an intermediate stage in fibril formation and its significance for tanning may essentially be disregarded. It is mentioned here by virtue of the fact that all computer modelling studies performed to date from a tanning chemistry point of view are based upon it. Its dimensions are given as a diameter of 3.8 nm and, depending upon the water content, an intermolecular distance of 1.2 ~ 1.7 nm, arranged in pentagon form. As the molecules exhibit virtually unlimited longitudinal growth, a dimension cannot be given for the length and must be selected at random for the purpose of calculation.

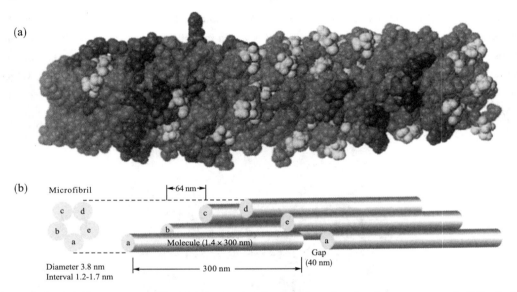

Figure 1.6 (a) Computer model and (b) structural principle and molecular arrangement of the Smith microfibril. Reprinted with permission from *Journal of the American Leather Chemists Association*, 1997, 92(8): 185-198.

1.2.3 The Collagen Fibrils

Since the late 1950s, the creation of the fibrils has been well understood. Under the influence of their charge pattern, the microfibrils are taken up laterally by each other, staggered by a quarter of their length up to an average between 100 nm and 130 nm, and grow longitudinally, apparently without limit, with the formation of a gap. The fibrils are accessible for characterization by common electron microscopic techniques, such as TEM, SEM and AFM.

The triple helices further assemble to form fibrils in which "gaps" are left longitudinally between the ends of the molecules. The gaps are 40 nm in length, their diameters vary according to the lateral molecule distance, i.e. from 1.2 nm to 1.7 nm. The arrangement of triple helices in a fibril is such that the N-termini of two axially adjacent triple helices are separated by $D \sim 67$

nm and the N-termini of two collaterally adjacent triple helices are separated axially by 0.54 D in the native hydrated state. This staggered arrangement creates alternating regions of low and high protein density along the fibril axis with a repeating unit of length D. A particularly common and important form of Type I collagen fibrils exhibits a nanoscopic signature, D-periodic gap/overlap spacing (Figure 1.7) [48-51]. It is found that the D-spacing differences arise primarily at the bundle level independent of species or tissue types such as dermis, tendon, and bone [52].

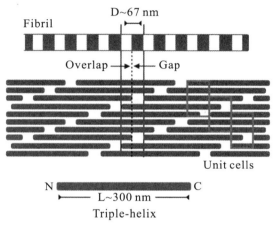

Figure 1.7 Schematic diagram of collagen fibril formation. Reprinted with permission from *Biophysical Journal*, 2016, 111(1): 50-56. Copyright (2012) Elsevier.

As the assembly is dependent upon the charge pattern, fibril formation can be influenced by modification of the latter. It was thus demonstrated some time ago that transformations on the basic groups involving tungsten triphosphate or on the acidic groups of the collagen involving uranyl compounds caused the typical cross-striation mentioned to be highly resolved and more sharply expressed by the "bringing into register" of the amino acids involved, i. e. of their lateral chains—an observation of interest also for tanning chemistry (Figure 1.8). Numerous other substances of physiological relevance and/or interest for the leather industry have been studied with regard to their influence upon fibril formation and the cross-striation periodicity [53, 54].

Figure 1.8   Fibril with cross-striation typical of collagen (AFM image)

1.2.4   The Collagen Fibers

The fibrils for their part continue to grow to form elementary fibers, fibril strands and fibers. The latter then form the fiber network which is of great relevance to the (mechanical) leather properties[55,56]. The arrangement of the fibrils can be seen clearly by electron microscopy in cross-sections of elementary fibers. The fibrils are bonded to each other not by cross-linking, but by intertwining (Figure 1.9). In contrast to the covalently cross-linked fibrils, the elementary fibers are intertwined only mechanically, an arrangement from which they derive their relative stability. They can be prepared from the collagenic fiber network as individuals. This method has been used for measurement of their dimensions by TEM: their mean diameter is 5 μm, their length up to several mm. The interfibrillar distance in the elementary fibers is between 115 nm and 155 nm. These dimensions refer however to the distance between axes, i. e. the extension of the fibrils themselves must be taken into account: "free" intervals, i. e. gaps which are actually accessible, exist only between the peripheral regions of the fibrils. They also vary substantially according to the water content. This constitutes a problem requiring further discussion. Fortunately, the excellent TEM images include images of elementary fibers which permit measurement of both their diameters and their distances.

Chapter 1 The Structure and Reactivity of Collagen

Figure 1.9 Elementary fibers: SEM images of grain (left) and flesh layer (right) of goatskin

Finally, the elementary fibers, with a wide range of fiber diameters, branchings and interwoven structures with the widest conceivable variation in weave angle, assemble to form the substantially thicker fibers and the fiber network as is familiar from the traditional view of a histological section provided by an optical microscope (Figure 1.10). The fiber thicknesses and the weave angle can be measured by optical microscope. They consequently vary, according to the skin types and the species, breed, age, etc. of the animal, within wide limits of approximately 140 μm and 200 μm. The interfiber distances likewise vary between 150 μm and 250 μm, and correspond to the macroscopic pores of the corium.

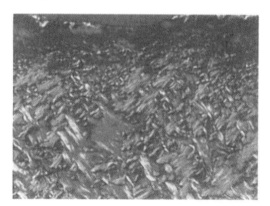

Figure 1.10 Collagen fibers in three-dimensional weave (polarized light image of a pigskin section)[1]

Collagen therefore consists of structural elements which form a structural hierarchy. Up to and including the fibrils, this structure constitutes the microstructure of the collagen; from the elementary fibers towards its macrostructure. The essential structural data are now known and can also be exploited for quantitative calculation of material transformations on the collagen, e. g. penetration and binding of tanning materials (Table 1.6). The structure of the collagen described above is in principle the same for all hide and skin species employed in leather manufacture, in particular as far as the fibrillar microstructure and reactivity are concerned. By contrast, clear differences exist regarding the macrostructure according to hide and skin species, origin, age,

Fundamental Collagen Chemistry in Leather Making

etc. The same applies to non-collagenous constituents, such as fat. To these are added topographical differences ("position") within the hide with regard to the weave density and angle of the collagen fibers; these differences also have a decisive bearing upon the mechanical leather properties. The stratigraphic differences between the individual hide layers, grain membrane, papillary and reticular layer are considerable.

Table 1.6 Data on the collagen structure

Structural element	Dimensions	Significance for leather properties
Amino acids	3 × 1052 amino acids per mol, of which 381 acidic (Asp, Glu), 268 basic (Arg, Lys, Hyl, His), 492 hydroxyl (Hyp, Ser, Thr, Hyl), and ~ 3 000 peptide groups	Center of reactivity, including water bonding
Molecules	1.4 × 280 nm, ~ 300 000 Dalton	See fibrils
Microfibrils	Φ 40 nm, 5 molecules	Metastable intermediate stage during fibril formation
Fibrils	Φ 100 ~ 200 nm, intra- and interfibrillar distance, dependent upon the water content, of up to 20 nm (capillary level) and fibril surfaces up to 30 m^2/g	Location of the intrafibrillar activity during transformation of the collagen into the leather matrix (including crosslinking)
Elementary fibers (fibril strands)	$\Phi \approx$ 2 mm	Location of the interfibrillar activity during transformation of the collagen into the leather matrix (including "coating")
Fibers	$\Phi \approx$ 100 mm, distances in the > 100 nm region (micropores)	Substantially responsible for garment comfort properties
Fiber networks	Φ in the mm region, woven, distances in the μm region (macropores)	Substantially responsible for the mechanical properties

1.3 How is Collagen Formed

Owing to its significance for cellular processes and the existence of numerous collagens with different functions, the biosynthesis of collagen has been a frequent subject of studies, and the domain of cellular biologists and biochemists. A brief summary of the synthesis of Type I collagen should therefore suffice. The genetic code for collagen is anchored at Position A 64 of the human genome. As the sections of the DNA responsible for collagen synthesis are known and sequence analysis of the nuclein bases is simpler than for the proteins. Collagen sequences are in fact now decoded in the main via DNA sequences and stored in suitable databases. Synthesis takes place in the connective tissue cells, the fibroblasts, beginning with the single-stranded α chains. The latter exhibit propeptides, at both the N- and the C- terminal end, which give the strands their solubility. Proline and lysin are not initially hydroxylated. Only in the extracellular space are these propeptides cleaved off, a part of the proline and lysin hydroxylated, the triple helix formed, and the crosslinkages developed which increase as the connective tissue ages. This progressively reduces the solubility of collagen and its ability to swell, while mechanical strength increases. These processes are all controlled enzymatically.

1.4 The Role of Water in Collagen

It is widely appreciated that water molecules play an invaluable role in governing the structure, stability, dynamics, and function of biomolecules. The hydration forces are responsible for packing and stabilization of the protein structure. Particularly, water participates in many hydrogen bond networks and screening electrostatic interactions[57,58]. Water is an integral constituent of the collagen structure. At the molecular level of the collagen structure, the water content acts as nothing less than a "crosslinker", stabilizing the helix structure[59,60]; further components are then responsible for the mobility of fibrils at the intrafibrillar and interfibrillar levels. In other words, water acts both as a tanning material and a softener. The distribution of water within the fibril controls their swelling behavior, shape, and mechanical properties on the nanometer scale[61]. Finally, the intercollagenous water is essential as a solvent for chemical transformations on the collagen. Reduction or for that matter removal of this water content has far-reaching consequences (Table 1.7).

Table 1.7 Type of water in collagen structure

Water "type"	Bond type	Quantity in g per gram of collagen	Quantity as a percentage of hide & leather	Effect when completely removed
Type I	Interhelically bound ("molecularly adsorbed")	0.01~0.07	to≈10%	Helix deformation
	Intrafibrillary bound ("molecularly bound")	0.07~0.25	to≈14%	(More extensive) Capillary contraction
Type II	Interfibrillary bound ("capillary bound")	0.25~0.50	to≈30%	Capillary contraction: volume/area shrinkage; isometric: strain incidence
Type III	(Loosely) bound in the fibrous structure ("bulk water")	0.50~2.0	to≈70%	Densification of the fiber network, pore contraction; isometric: strain incidence, lack of solvent for process chemicals

Consideration of this "system character" of collagen and water is crucial to an understanding of many properties of collagen and leather[62]. It should suffice here to point out that the thermal stability of collagen and leather is a function of its water content. The thermal stability of water-free collagen is approximately 200 ℃ and is reduced somewhat by tanning. The hydrothermal stability of fully hydrated skin collagen is known to be only 60 ℃, and is increased by vegetable and aldehyde tanning by around 20 ℃ and by chrome tanning by up to 50 ℃.

1.5 The Reactivity of Collagen

As a consequence of the different functional groups which it contains, collagen is capable of numerous transformations: it is reactive. Were this not the case, tanning and other material transformations on collagen, and therefore leather manufacture, would not be possible. The blocking, transformation or introduction by collagen of further functional groups is however a further important aspect of its reactivity. Owing to the comprehensive information available on collagen, the modes of transformation relevant to leather manufacture of the individual functional groups with the help of the chemically widely disparate tanning agents are known for the greater part (Table 1.8)[63].

Table 1.8 The potential reactions on collagen during leather-making

Functionality	Bond type	Typical tanning agents
Carboxyl groups	Complex bonding	Metallic salts, in particular basic Cr(Ⅲ) sulphates
Basic groups	Covalent bonding	Aldehydes, diisocyanates, etc.
Peptide groups	Hydrogen bonds	Phenolic natural and synthetic tanning agents
Surface, overall	Hydrophobic, "van der Waals" bonds	Including polymers, tensides
Pores/capillaries	"Fillers"	Various substances

The various means listed below for the binding of tanning material can be substantiated by independent evidence, as has long been the case in particular for the chrome tanning process with its great industrial significance but are also suitable for new tanning materials (Table 1.9).

Chapter 1 The Structure and Reactivity of Collagen

Table 1.9 Demonstration of chromium binding to the carboxyl groups of the collagen

Substrates and/or modification	Effect, proof of Cr – (—COOH) – interaction
Chrome tanned collagen	pH-dependence and level of chrome bonding
	Shift of isoelectric point (IEP) towards higher pH values
	Change of FT-IR and ^{13}C NMR spectra
Deamidation by liming	Increase of chrome binding
Blocking of —COOH groups by methylation/esterification	Decrease of chrome binding
Introduction of additional —COOH groups by succylination, reaction with amino acids and aldehydes, glyoxalic acid, etc.	Increase of chrome binding
Electron microscopy	Highly resolved crosss-triation by chrome staining in accordance with amino acid sequence
Collagenase degradation	Enrichment of chrome in —COOH cluster peptides
Model peptides	Chrome binding depends on the content of —COOH groups
Polyamides	Almost no chrome binding
Ion exchanger	Chrome binding depends upon content of —COOH groups

At present, a need is apparent for further development in traditional tanning theory, not least with regard to greater consideration to the role played by water in the binding of tanning material, cluster formation, etc[64,65]. This aspect will also have consequences for future search strategies for new tanning materials.

1.6 The Nature of Leather Formation and the Characteristics of Leather

Based upon current knowledge of the structure and reactivity of collagen, the nature of leather formation is now well understood[66-71]. In view of the works published on the subject, the technology of leather manufacture will not be discussed in detail at this point. It need only be said that when hides and skins have been unhaired and fleshed and the resulting corium relieved of accompanying, non-collagenous substances such as globular proteins, proteoglycans and fats by the beamhouse processes, the pelt is left which comprises approximately 30% Type I collagen and 70% water.

The nature of leather formation is revealed by a comparison between the process of drying of the pelt transformed into the leather matrix on the one hand and "mere" drying out to form a parchmentlike material on the other:

Parchment ← Pelt → Leather matrix

In the first case, the collagen undergoes a series of property changes which are typical of completion of leather formation. At the same time, the material acquires important handling and processing properties which can be assigned to categories. The "basic properties" are those characteristics of the essence of leather formation, as follows:

- The "permanent leatherlike drying-out" as described by Gustavson.
- The enhanced hydrothermal stability.
- The reduced water take-up (*bulk water*) and swelling capacity.
- The resistance to microbial (enzymatic) attack.

1.6.1 Leatherlike Drying Out

The "permanent", water-resistant, leatherlike and porous drying-out as described by Gustavson is without doubt the most important basic property of the leather and is responsible for the majority of its physical characteristics. It is responsible for the difference between parchment and leather[72-74], and is measured in terms of the leather yield (Figure 1.11). The difference between the agglutination of fibers in parchment and their isolation in leather can be demonstrated by AFM (Figure 1.12).

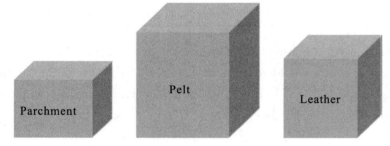

Figure 1.11 Diagram showing the volumetric change during drying associated with transformation from pelt to parchment and leather

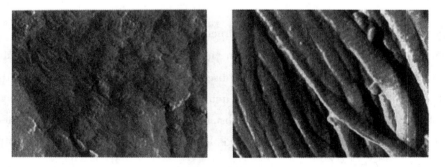

Figure 1.12 Parchmentlike and leatherlike drying out (AFM × 50000)

In parchmentlike (hard) drying-out, the flexible structural elements of the collagen are progressively drawn towards each other under the influence of the surface tension of the water during dehydration, ultimately bonding to each other due to their reactive groups, thus leading to deterioration of the hierarchical structure[75]. Collagen (from the Greek for "glue") does justice to its name. If the water is replaced by a water-soluble solvent or supercritical CO_2, the contracting effect exerted by the surface tension of the water is no longer present; at the same time, dehydration causes intrafibrillar compression of the collagen structure resulting in a stiffening of the fibrils. As a result, the fibrils lose their flexibility, remain isolated and fail to bond; a very leatherlike (soft) drying-out of the higher-level fiber composite is the result. Pelt dehydrated in this manner is not true leather, however, as it is not water-resistant. If, however, e. g. fats are added to the solvents (150 years ago, Knapp used stearic acid for this purpose; 100 years later, Heidemann and Riess used silicons). The fibrils and fibers are coated with a hydrophobic film, and leatherlike drying-out occurs repeatedly even following renewed exposure to water. This effect may be enhanced by gentle flexing and stretching. Leather has been produced by hydrophobic coating.

In the normal leather production process, the stiffening of the fibrils required for leatherlike drying-out is achieved by water-resistant tanning, and the softness-inducing mobility of the fibrils and fibers and the permanence of the "leatherlike drying-out" are promoted by fatliquoring (hydrophobic coating) and by the mechanical processes of staking and milling. The stiffening effect of tanning is based partly on crosslinking of the collagen at the intrafibrillar level; the effect is reflected in the increase in the T_s. This crosslinking process may take place fairly locally, or over a wider area according to the clusters already referred to on reactive groups of the collagen as a function of the type and quantity of tanning material employed (chromium, R—CHO etc. in the former case, phenolic tanning materials etc. in the latter). In addition to the crosslinking function, the "distance-enhancing", filling effect of the majority of tanning materials is also important (Figure 1.13). In addition to leatherlike drying-out, the transformation of collagen into the leather matrix causes certain other changes in the collagen's response to environmental influences.

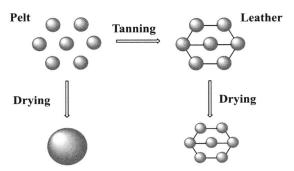

Figure 1.13 Diagram of the nature of leather formation by which leatherlike drying-out is ensured by the crosslinking and distance-enhancing action of the tanning agents

1.6.2 The Hydrothermal Stability (Shrinkage Temperature T_s)

The increase in T_s is, as already stated, a consequence of crosslinking of the collagen. It reflects the thermal resistance of fully water-saturated leather, i. e. the hydrothermal stability,

Fundamental Collagen Chemistry in Leather Making

which is by no means a typical application, with the exception of heat setting and certain other special cases. Leathers with the widest possible range of T_s, even "pseudo leathers" of unchanged T_s, such as acetone-dehydrated, silicon-impregnated pelts mentioned above, have a practical value of their own. The search for new tanning agents must allow for the fact that a high T_s is not necessarily one of the objectives.

1.6.3 Water Take-up Swelling Capacity

The reduced swelling capacity of the collagen resulting from "crosslinking" as a function of the tanning type and intensity of the collagen substantially reduces its water take-up. Reduced further by fatliquoring, filling and hydrophobation, this reduction in the leather matrix's water take-up compared to the original collagen is a basic property of definite practical value. The superior mechanical dewaterability of leather compared with the pelt is the very characteristic which permits shaving. The reduced water take-up and swelling is limited almost entirely to the water (Type III) contained in the pores and microcapillaries. The bound water (Types I ~ II) is largely unaffected by this mechanism and fluctuates with native collagen as a function of the temperature and relative humidity with all its consequences for the leather properties. The swelling capacity of the tanned fibrils has, however, by no means been lost. On the contrary, environmental scanning electron microscope (ESEM) images reveal how the fine structures of the leather matrix change as a function of the water content.

1.6.4 Resistance to Microbial (Enzymatic) and Hydrolytic Attack

A typical feature of leather has always been its greater resistance to microbial (enzymatic) attack compared with that of pelt/hide. This increased resistance must be considered critically, however. The water content is the first deciding factor: even untanned hide is resistant to microbial attack if the water content is sufficiently reduced (drying, salting of raw hide; parchment). The fact that it decomposes more rapidly when wet than wet leather is due firstly to its numerous soluble constituents, which constitute an ideal breeding-ground for bacteria, and secondly to its higher level of swelling (see above). The microbial decomposition, for example in the composting test, of pure collagen subjected to different tanning processes differs only in its speed, as a function of the level of crosslinking and the possible biocidic action of ingredients released from the leather; the action as such is no different in principle. The latest findings concerning the effect, as a function of the tanning type and intensity, of collagenase, an enzyme which particularly breaks down collagen, are particularly interesting in this regard[76].

1.6.5 The Handling Properties of Leather

Leather continues to enjoy consumer popularity due to its handling properties, the origin of which can be found largely in the structure and reactivity of the collagen. The popularity begins with the aesthetic features of the attractive surface and pleasant feel, all-important properties of water vapor absorption and permeability (respiration activity) with their importance for garment comfort, and not least its mechanical properties, such as optimum stress-strain characteristic due to the netlike deformation of the fiber network under low load (shape adjustment) on the one hand, and resistance to (major) deformation on the other (shape retention). These characteristics, which are inherent to the collagen, can be controlled within quite wide limits by the processes of leather manufacture with reference to the intended use of the leather.

1.7 Concluding Remarks

We began with a generally product-oriented definition of the material "leather", and have concluded with an attempt to define leather in terms of the structure and reactivity of the collagen:

Leather is a material, derived from Type I collagen of the corium of hides and skins, the reactivity of which has been used for a range of different chemical transformations in order to prevent the collagen structure from collapsing during water-resistant drying and thus to keep the individual collagenous structural elements mobile. This is achieved by stiffening and/or coating of the collagen fibrils and fibers by the substances acting upon them. The transformation of the collagen into leather is associated with an increase in the hydrothermal stability, reduced swelling capacity, and increased resistance to microbial/enzymatic attack.

Leather production is an ancient process. To develop it further, to increase the practical value of leather, to counter ecological and economic constraints effectively, remains a challenge of our time. To master these aspects will fulfil an expectation which can be formulated as:

"*Leather—a material with a history, a material with a future.*"

Chapter 2 Physical and Chemical Processes on Collagen and Its Transformation into the Leather Matrix

2.1 Introduction

Models were produced and model calculations performed in order to ascertain what quantities of substances may interact under what conditions with the collagen: "interaction equivalents". The resulting material bond may be the result of chemical/stoichiometric interactions: "bond equivalents". In addition, the different secondary valence bonds (hydrogen bonds, dipole linkages, van der Waals' and hydrophobic bonds) give rise to secondary valence attachment on available surfaces: "attachment equivalents". Finally, the interactions may also lead to attachment by filling in available cavities: "interstitial deposition equivalents".

The objective of the present study was to describe these various interactions and to quantify as far as possible the three different interaction equivalents. The calculations performed for this purpose are based upon the structural hierarchy of collagen, i. e. they consider the chemical aspect of amino acid composition and sequence, and the structural aspect of the accessible molecule surfaces, microfibrils, fibrils, elementary fibers and fiber surfaces and free cavities present between them (nano-, micro- and macro- pores). The pore and capillary structure in the collagen, i. e. voids, can be determined from the distances between the individual collagenous structural elements (CSEs). The voids are a requirement for penetration by the substances reacting with the collagen. It is common for these voids to be classified as pores and capillaries, and possibly distinguished even further, according to their dimensions. This classification is not always consistent. It is therefore better for the dimensions of the voids to be assigned to the individual CSE levels (Table 2.1). Certain selected, typical industrial low-molecular electrolytes, tanning materials, dyes and fats of varying relative molecular weight and molar quantity were considered as examples of reacting substances, as were water and nitrogen, the latter owing to its significance for the measurement of specific surface areas on collagen and leather by means of the BET method (based on the Brunauer, Emmett and Teller model)[77]. Leather processing involves diffusion of various chemicals through the pores of collagen fiber matrix[78,79]. Analysis of pore structure of the collagen is important to understand process of diffusion and adsorption involved during the leather matrix formation as well as characterization of comprehensive properties of the resulting leather[80,81].

Chapter 2 Physical and Chemical Processes on Collagen and Its Transformation into the Leather Matrix

A number of methods are available for measurement of the pore size and distribution. These range from BET measurement, through feature analysis by means of SEM, Hg porosimetry, and capillary flow porosimetry technique[82, 83]. Metrological aspects will not be discussed here, but the fundamental pore distribution is shown by way of example based upon the conventional procedure using Hg porosimetry (Figure 2.1). The majority of voids are shown to be on the elementary fiber and fiber level, which is an essential aspect for consideration of the interstitial deposition equivalents. Attention is also drawn in this context to the fact that the pore and capillary radii are dependent upon the state or level of swelling of the collagen, and probably also upon the intensity with which the hide is opened up. The first of these phenomena, at least, has been demonstrated for intrafibrillar (i. e. intermolecular) distances by X-ray structural studies.

Table 2.1 Pore and capillary structure of Type I collagen

Description	Structural aspects	Dimension
Total voids, expressing the overall porosity	Voids between all collagenous structural elements (CSEs)	Porosity in %
Nanopores	Distances between helixes in the molecule (a) and intermolecular distances within the microfibrils and fibrils (b)	(a) 0.15 nm; (b) Up to 1.7 nm
Micropores (capillaries)/semipores	Distances between fibrils in the elementary fibers	0.1 μm (100 nm)
Pores	Distances between elementary fibers	1 ~ 3 μm
Macropores	Voids/distances between the fibers in the fiber network	>3 μm

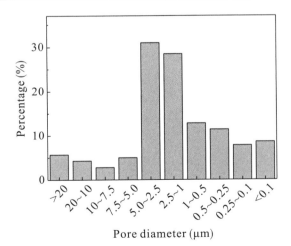

Figure 2.1 Pore distribution in acetone-dehydrated pelt determined by Hg Porosimetry

The modes by which the individual substances may reach their location of activity in the collagen were considered. The diameter of the diffusion paths represents an important factor here ("diffusion funnel"). The various physical and chemical attractive forces and the charge characteristics on the surface of the collagenous structural elements, which can be modified by the substance binding and thus influence the binding of further substances, are also of great significance.

The results were evaluated with regard to tanning theory, in consideration of product development strategies, and with regard to the need for further and more specific studies. In the light of the breadth of processes occurring on collagen, by no means all questions have been answered thus far. The present study, which is principally methodical in its orientation, is intended to prompt discussion and further research, and perhaps production of a computer-based model of the processes occurring on collagen.

2.2 Preliminary Remarks

The objective of the present study is to aid understanding of the physical processes arising during the structural transformation of collagen into the leather matrix, and in the process to facilitate the search for new agents, specifically compounds which stabilize or soften the structure. Based upon current knowledge of collagen and the substances reacting with it, basic assumptions are postulated in the interests of simplicity and industrial relevance. Importance has been attached to a graphic presentation.

The main text will first address certain principles of tanning theory, the initial assumptions and definitions, and the procedure. The question of how the active substances penetrate the collagen and its reactive centers will then be considered, and what quantities of substances can be bound by the collagen, specifically as a summary of the interaction equivalents, categorized: chemical aspects (bond equivalents) in terms of the number and accessibility of functional groups of the collagen and of the reactants; spatial aspects (attachment equivalents) in terms of the accessible surfaces of the individual structural elements of the collagen; steric aspects (interstitial deposition equivalents) in terms of the accessible pores and capillaries.

2.3 Principles of Tanning Theory, Basic Assumptions, Definitions and Methods

The transformation of hide into leather is a multistage process of physical and structural modification of the collagen. The part of the hide relevant to leather, the corium (cutis, dermis), is separated from the two other layers of the hide, the epidermis (the upper layer of the hide, together with the hair) and the subcutis (hypodermis), stripped as far as possible of its non-collagenous components and its chemistry and structure prepared in the process for tanning ("opening up of the hide"). The latter process promotes accessibility to the reactive centers in the microstructure of the collagen, and liberates further —COOH groups by deamidation. The

Chapter 2 Physical and Chemical Processes on Collagen and Its Transformation into the Leather Matrix

main function of the process, however, is one of cleaning. The processes described are performed in the beamhouse and are completed at the washing stage following deliming/bating.

For the purpose of the present study, the pelt collagen retained by this process is the collagen which is transformed into the leather matrix. All chemical and structural aspects described here relate to this collagen.

The leather matrix refers to the integral system of collagen, ingredients of various kinds bound to the collagen, and water, which is responsible for the leather properties as a function of the intended purpose of the leather, with the exception of the surface treatment (finishing). The latter does not fall within the scope of the present study.

The typical composition of pelt collagen, that of the parchment obtained from it by mere drying, and that of damp and dry leather matrix, is shown in the diagram by a mass balance of a chrome-tanned upper shoe leather. Wet-white automotive leathers differ only slightly in composition (Figure 2.2).

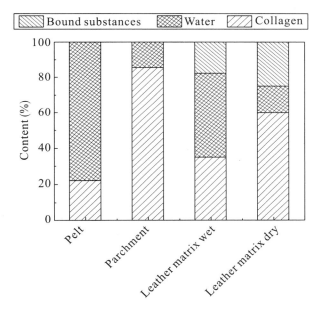

Figure 2.2 Composition of pelt, parchment and wet and dry chrome tanned leather matrix

2.3.1 Principles of Tanning Theory

In conjunction with the discourse on possible innovations in leather, a comprehensive literature survey was performed of previous studies of leather theory. A comprehensive presentation is available. For the present study, it is therefore sufficient to remind the reader of certain basic principles.

The transformation of collagen into the leather matrix means the taking of suitable measures during dehydration in order to guarantee a soft, "leatherlike drying-out", in complete contrast to the "hornlike, transparent" drying-out of parchment. In physical terms, this means the assurance of a suitable porosity, a consequence of isolation of the collagenous structural elements. The varying yields of pelt, parchment and leather matrix provide an illustration of this aspect (Figure 1.11). Incidentally, the yield is normally expressed in cm^3 leather/100 g collagen. Typical values are 170 ~ 220 for chrome-tanned leather, 250 ~ 400 for vegetable-tanned leather and 90 ~ 120 cm^3/100 g collagen for parchment. The yield on leather volume, cm^3 leather/cm^3 of wet pelt, the porosity in %, and finally, the apparent density in g/cm^3 are other numerical expressions for the same property of CSE isolation by leather formation. This deciding characteristic of leather must withstand practical exposure to an appropriate influence of water without undergoing substantial change.

The transformation defined of collagen into the leather matrix is brought about by the combination of tanning processes performed in several stages (pretannage, main tannage, retannage), supported by fatliquoring and (to a limited extent) by dyeing, influenced by the drying conditions, and promoted by the mechanical processes, particularly those of staking and milling. The structural changes to the collagen are those responsible for leather formation, brought about essentially but not exclusively by physical transformation processes on the collagen.

The transformation of collagen into the leather matrix is associated with other changes in properties. These include primarily the modified hydrothermal (tested as T_s) and thermal stabilities (measured by DSC), the enhanced resistance to microbial attack, the modified swell capacity, and other properties. Despite the significance of these characteristics, they are not essential to the present study.

The definition of leather formation selected excludes procedures, which although bringing about a leatherlike drying-out, involve dehydration not by drying, but by solvent treatment (with acetone, alcohol, supercritical CO_2) or freeze-drying, and attain the water-resisting property, initially absent, only through additional impregnation. They are of considerable interest in tanning theory for model studies and screening, and also of great practical interest, but lie beyond the scope of the present study.

2.3.2 Assumptions and Definitions

Type I pelt collagen is that considered throughout the present study. It is regarded here as a system comprising Type I collagen and water. The low content of other collagens and proteoglycans and the residual fat content of pelt will not be considered here. Where the study refers to industrial data, such data refer to pelt collagen from cattle hides. The fiber-forming Type I collagen is characterized by its hierarchical composition from several different structural elements (molecules, microfibrils, fibrils, elementary fibers, fibers, and fiber network). These collagenous structural elements are referred to in the present study as "CSEs" (Figure 2.3).

Chapter 2 Physical and Chemical Processes on Collagen and Its Transformation into the Leather Matrix

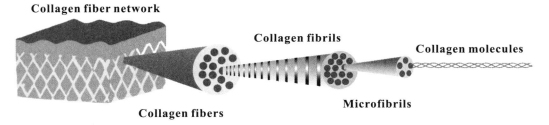

Figure 2.3 Principle of the structural composition of Type I collagen

Attention is expressly drawn to the following aspect: precise structural data exist only for the Type I collagen molecule with the composition $[\alpha 1\,(I)]_2\,[\alpha 2\,(I)]$. This data applies to all Type I hide collagen molecules, irrespective of the animal species, breed, origin, etc. and the point of recovery from the hide. The same applies to the postulated microfibrils formed metastably from five molecules ("Smith" microfibrils). The latter's composition is, however, not entirely undisputed[84]. They represent only an intermediate stage of the fibril growth and are in fact not relevant to transformation of collagen into the leather matrix. They (or a model of them) can however be used to illustrate effectively the principle of composition of the fibrils, the consequence of transversely and laterally staggered molecule assembly and certain key structural characteristics. Segments of the microfibrils have also been the repeated subject of computer models of the tanning processes—a further reason to consider them among the CSEs (Figure 2.4)[85-87].

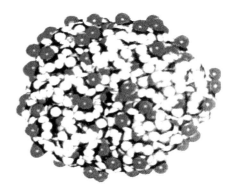

Figure 2.4 Cross-section of the computer model of a microfibril

Conversely, the structural data for the fibrils, the elementary fibers, the fibers and the fiber network depend to a large degree upon the influencing factors stated. In that respect, numerical data, for example, for the diameter, for instance of fibrils and fibers, represent only statistical mean values of a considerable dispersion. The same then also applies logically to the radii of the capillaries and pores. It was not possible to take into account the well-known topographic and stratigraphic differences in the macroscopic collagen structure in the present study; these must however be considered in experimental studies.

The rule which may be formulated from the foregoing is that the greater the extent to which processes on the collagen take place in its microstructure, i. e. in effect at molecular level, the more precisely they are (can be) described. The restrictions stated with regard to the structural data do not apply to the chemical characteristics: the amino acid composition and sequence are specific to Type I collagen and independent of its origin and spatial structure.

As far as the substances involved are concerned, selected typical examples of the main tanning agents and fillers and a small selection of dyes and fatliquoring agents were included in the study in addition to N_2 (test gas employed for measurement of the specific surface area by means of the BET method), water and certain low-molecular electrolytes. N_2 is generally employed as an inert gas for the investigation of specific ("inner") surface areas by means of BET measurement. All measurements on collagen and leather show that even on a pelt collagen with a density of <0.5 g/cm^3, i. e. a porosity of $>60\%$ and a void volume of >1.6 cm^3/g collagen, dehydrated with acetone and supercritical CO_2, a specific surface area of only approximately $25 \sim 30$ m^2/g of collagen is measured. With reference to the surface areas of the individual CSEs, this corresponds to coverage of the fibril surface. This means that N_2 molecules, despite their low dimension of 0.02 nm^3, do not penetrate the inside of the fibrils, even though sufficient space is theoretically available for them. The collagen molecules "stick" within the fibrils either during dehydration in the manner of concentric shrinkage and/or in preparation for the BET measurement (high vacuum) itself.

Leathers possess a specific surface area of $0.5 \sim 5$ m^2/g, i. e. only the fibers themselves are accessible to the N_2. Either the lower-level diffusion paths are blocked even to N_2 by tanning materials, fats and fillers, or the collagen structure undergoes further densification during drying of the leather. The higher apparent density, the reduced porosity, and not least the low specific surface area of parchment (mere dried-out pelt collagen) suggest the latter.

Chemically bound quantities, i. e. the bond equivalents in the narrow sense of the term, were calculated from stoichiometric equivalences, most simply and accurately in the case of electrostatic interaction (ionic bonding). In the case of covalence, complex and hydrogen bonding, calculations were performed for variants under certain defined conditions (e. g. single, two and multi-point bindings, limitation of the functional groups considered, etc.). Coordinative secondary valence bonds (dipole attraction, van der Waals' and hydrophobic bonds) were not studied in greater detail at this stage, although they doubtless represent a worthwhile field of further study, not least with regard to the relationship between structural characteristics of a great many substances.

The steric calculation of the attachment equivalents referred to in the introduction was based upon the molecular morphology of the substances involved (cubic, rectangular) and their possible position on the collagen surface, parallel or perpendicular to it (Figure 2.5).

Chapter 2 Physical and Chemical Processes on Collagen and Its Transformation into the Leather Matrix

Figure 2.5 Model of the substance attachments on the collagen surface[1]

These models were also employed for assessment of the scope for diffusion and estimation of the interstitial deposition equivalents in the various voids of the collagen. As far as the substances reacting with the collagen are concerned, recourse had to be made in many cases to approximated structural data and models.

The section below discusses how the active substances enter the collagen. It therefore concerns the diffusion and penetration of the active substances at the actual centers of reaction and bonding of the collagen. The laws governing these processes have been widely described and studied, but often in other contexts, or from a non-universal perspective. A brief description from the point of view of the present study was therefore deemed appropriate.

2.4 Diffusion and Penetration

The questions posed here concern the forces triggering and promoting diffusion, the length of the diffusion paths, and the relationship between the diameters of the diffusion paths and the particle sizes of the diffusing substances. Means for accelerating diffusion and for the generation of higher molecular and/or more highly aggregated particles *in-situ*, i.e. once the substances have penetrated the collagen microstructure or macrostructure, are also the subject of consideration.

2.4.1 Laws of Diffusion

Pelt consists of approximately 25% collagen and 75% water; this corresponds to about 300% water in terms of collagen. Of this water content, approximately 10% is bound to a greater or lesser degree to the collagen; the greater part of the water is "free" ("*bulk water*"), and consequently acts as a solvent, thus forming the "inner float". This contrasts with the "outer float", whether in a pit, paddle, drum, mixer, or other tanning vessel in which the reacting substances are initially present molecularly dispersed or in colloid suspension, and also suspended in agitated systems. In the simplest case, that of a static system containing substances with no affinity for collagen, the concentration gradient present gives rise initially to diffusion from the outer to the inner float in accordance with Fick's Laws. Simple stirring, the movement of a paddle wheel or movement of the drum accelerate this diffusion process by more rapid equalization of the concentration in the outer float. Ultrasound, for example, is subject to the same mechanism.

The situation is different where substances with an affinity for collagen are concerned. In this case, diffusion is accelerated considerably by constant substance adsorption as a result of disturbance of the diffusion equilibrium. The essential mechanism is that of Langmuir adsorption. Initially, it is of secondary importance which bond types and which functional groups of the coreactants are responsible for the affinity; as far as the collagen is concerned, the most important factor is their accessibility, i. e. the dimensions of the diffusion paths and the available specific surface area of the collagen.

If a hydrogen ion H_2O^+ is considered which is to penetrate the interior of the pelt up to a molecularly bonded carboxyl group, the path of some 3 mm to the middle of the pelt may equally appear either short or long. This distance may be put into perspective by making the actual dimensions of the collagen molecule of 300 nm × 1.4 nm equal to 1 mm. The diffusion path to the inside of the hide then equates to no less than 1 km! In the processes, the collagen molecule must pass through the various pore and capillary radii in the μm to the nm range. In the process, it negotiates a "diffusion funnel" which serves as an illustration of the diffusion paths (Figure 2.6).

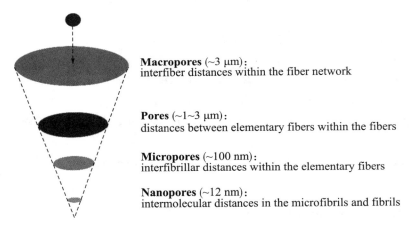

Macropores (~3 μm): interfiber distances within the fiber network

Pores (~1~3 μm): distances between elementary fibers within the fibers

Micropores (~100 nm): interfibrillar distances within the elementary fibers

Nanopores (~12 nm): intermolecular distances in the microfibrils and fibrils

Figure 2.6 "Diffusion funnel" of collagen[1]

The generally accepted relationship for the diffusion of vegetable tanning agents may serve here as an example of the laws governing penetration. According to these laws, the penetration depth is dependent upon a diffusion coefficient multiplied by the square root of the tanning agent concentration and the duration of tanning, the diffusion coefficient varying according to tanning agent type. As soon as the diffusing substance reaches the inner float from the outer float, it is taken up by the collagen. This logically takes place, with a time lag, from the outer layers towards the inside of the pelt ("active exhaustion"). Provided the supply of the substance remains below the saturation limit of the collagen, its bonding capacity for the substance concerned, this exhaustion process will repeatedly restore the concentration gradient between inner and outer floats, and diffusion will continue, in the best case resulting in 100% exhaustion of both floats.

With astringent, larger-particle substances of high affinity, bonding of the substance initially on the outer layers of the pelt may lead to blockage of the diffusion paths. This development may also be promoted by the fact that any substance binding constrains the diffusion paths, not least owing to the dehydration associated with the binding of the substance ("over-tanning"). In this context, it must be borne in mind that all tanning agents which owe their affinity for collagen to their ability to form hydrogen bonds, their dipole character, or their "secondary valence activity" in general, also aggregate to form larger particles (vegetable tanning agents, phenolic replacement tannins, substantive dyes, etc.). This aggregation varies according to the concentration and may result in particle sizes in the μm range and (apparent) relative molar masses of several kD. Michailow has developed interesting model concepts ("osmotic cell") in this regard which have attracted little attention to date, but which deserve closer consideration with reference to the original literature.

Great attention is paid in practice to the beginning of tanning with substances of small average particle size distribution, a "golden rule" of tanning. Examples cited here are the countercurrent principle in the suspenders, *in-situ* particle enlargement by basification of the chrome tanning material following penetration, and the creation of tanning material as such *in-situ*, for example during two-bath chrome tanning. *In-situ* tanning material creation encompasses all procedures in which the action of a primary component is enhanced substantially by after-charging with a second. The increase in T_s of vegetable-tanned leather achieved by pre-treatment or post-treatment with aluminum or other metallic salts, oxazolidines, CH_2O melamine resins and CH_2O protein condensates are particularly familiar examples here[88]. The "condensation tannages" (polyvalent phenols, aluminum salts, formaldehyde) are worthy of mention in this context.

The "synergetic" effects of such combinations of substances were explained by the concept of the combination of compounds with dual affinity proposed by the author. The most recent work in tanning theory performed by Covington et al. postulates an interplay of tanning material(s) and water whereby supramolecular structures (matrices) are formed around the collagen which are decisive for stabilization of the structure and for the increase in T_s[89].

Attention is also drawn to the fact that the exploitation, by compounds used in after-charging, of residual valences of tanning materials which have already been bound has for a long time been state of the art, particularly in chrome tanning: bonding of —COOH functionalized polymers, dye bonding, copper stearate formation, etc.

The dispersion of larger-particle substances, e. g. vegetable tanning agents by synthetic organic tannin materials, is another means of facilitating and/or accelerating diffusion. Any binding of a substance also stabilizes the collagen structure: properly performed pretannage processes thus substantially facilitate and accelerate subsequent main tannage processes, which can then be performed "more robustly" with regard to the particle size. Conditioning processes, e. g. by means of pickle or mere salting, are similar in their effect. This relationship between structure stabilization and pore constriction is the very area in which many combination tannage processes present risks and offer opportunities which are worthy of closer consideration.

Diffusion within the inner float is promoted by the drum movement. If additional temperature effects are disregarded, any increase in temperature is known to increase the diffusion speed. The flexing of the pelts brought about by the drum movement is thus the principal cause of movement of the inner float and therefore constant renewal of the diffusion gradient. This drum effect is, however, brought to bear only if the collagen is not strained by swelling.

It is evident that for a given substance supply in relation to the collagen, the substance will be exploited more quickly and completely the lower ("shorter") outer float. This follows by necessity from the higher concentration in shorter floats. In many cases, however, the dependence of the particle sizes upon the concentration is a further factor. In the case of chrome tanning materials, dilution promotes hydrolysis, in turn promoting an increase in particle size; tanning in high-concentration floats, in its extreme form in undissolved chrome technology, is therefore beneficial for chrome tanning. Here too, *in-situ* formation of tanning material by the lifting of temporary masking is a factor. The exact opposite applies in the case of vegetable tanning materials: an increase in the concentration promotes aggregation with a consequent increase in particle size (see above). Combination with dispersive organic tanning materials permits low-float or non-float processes, however.

For the sake of correctness, it should be noted that the terms "floatless" and "short-float" as they are frequently employed always refer only to the outer float. Owing to the water released from the pelt during tanning/fixing and to osmotic effects in general, this only describes the initial state, however.

As the float is not usually 100% exhausted, the measured ratio between the inner and outer floats and the substance concentration in the outer (residual) float permits calculation of the unbound substance components present in the inner float. In the absence of downstream fixation processes, these are then the "mobile" components present in the leather, i. e. they can be washed out. In certain cases, however, it is possible to immobilize such substance components, for instance by means of precipitation induced by a change in pH value.

2.5 Material Binding in and on the Collagen

Firstly, it must be remembered that the different substances differ in their affinity and astringency and in their chemical structure, and consequently in their reaction with the bonding center of collagen. These differences dictate the bond quantities and strengths; however, they also influence the binding speed. Consideration must also be given to temporarily dispersed or solubilized systems which separate the dispersed components insolubly upon withdrawal of the stabilizer, e. g. the emulsifier in the case of fatliquor. Systems sensitive to pH which respond in a

Chapter 2   Physical and Chemical Processes on Collagen and Its Transformation into the Leather Matrix

similar way have already been alluded to. Substances separated in this way by destabilization of the system (i. e. by variation of the parameters) then remain in the leather following drying in the same way as "properly" bound substances, although they may have been separated "only" as insoluble filler. The arithmetical models for the interaction equivalents defined here, i. e. the (approximately stoichiometric) bond equivalents, "attachment equivalents" and/or "interstitial deposition equivalents", thus follow from these principles.

Diffusion/penetration are generally regarded as the variables determining the duration of transformation of collagen into the leather matrix. The prevailing chrome tannage is, however, the classic example of how fixation processes may indeed be of long duration. These "post-tanning processes" are familiar to experts in the field, particularly for chrome tannage, and have also been investigated to some degree scientifically; they require more detailed research, however, particularly with regard to the use of new substance systems.

At this point, certain remarks are called for regarding the influence of the charge characteristics of the collagen and the leather matrix upon the diffusion and fixation processes. Ion charges have approximately 20 times the range of dipole charges and similar secondary valence forces due to their electrostatic nature, and thus considerably promote the rate of take-up of ions of opposite charge. In addition, the charge character has a general determining influence upon take-up and fixation of the active substances by the collagen, or by the (partly preformed) leather matrix during successive processes according to the principle of attraction of opposite and repulsion of identical charges. It is thanks to G. Otto, BASF, that this concept was applied to tanning chemistry with greater relevance to practical application in the 1950s, and it has been taken up by more recent studies, such as those of Heidemann. Knowledge of the influences of charges upon the processes occurring on the collagen have become textbook material as far as the qualitative aspects are concerned. The situation is different as far as quantitative relationships and the deeper theoretical treatment are concerned. Much work remains to be done on the measurement of particle charges of the active substances and the charge pattern of the collagen. Even basic IEP measurement or the measurement of Zeta potentials are far from simple.

The stability in water of the leatherlike drying-out and thus of tanning material fixation is by definition the characteristic of leather. This has its limitations, too, however; many of the fixations of (tanning) material are known to be reversible under changes (in some cases minor) in pH, the influence of complexing agents, or sustained exposure to water. Should a piece of wet blue be drummed sufficiently long in water with a piece of pelt, both pieces will ultimately be "half" chrome-tanned. For water to be taken up at all by the leather matrix and to be able to exert its detrimental effect, a certain hydrophilicity of the collagen and porosity and capillarity of the leather matrix are required. The majority of tannages do not change the hydrophilicity and preservation of the pore structure is the very essence of tanning. The hydrophobing action of fatliquoring thus continues to be essential besides its softening action.

Provided it does not impair the practical value of the leather, a low bond strength under exposure to water can to some degree be justified. Hardening following exposure to water can be eliminated by suitable care (e. g. of shoes) or mechanical stretching (e. g. gloves, wash leather). These aspects must, however, be considered during the search for new tanning materials, and attention must be paid in the search strategies to systems, in the form of water-insoluble compounds which are generated *in-situ* or separated by destabilization, which cause leatherlike drying-out. Should they also be hydrophobic, all the better.

At this point, brief mention will be made of fatliquoring. Fatliquoring is performed essentially for two reasons. Firstly, the water take-up discussed above is reduced (the aspect of hydrophobization). Encapsulation in fat certainly also delays water-related impairment of the tanning material fixation. Secondly, however, the fat contributes to leatherlike, soft drying-out of the leather matrix. A function which in fact by definition lies within the area of tanning. A direct relationship does in fact exist between the type and intensity of tannage on the one hand and the need for fatliquoring on the other. Modern chrome-free tanned automotive leathers are a perfect example of how leather having undergone "meagre" tanning can be given the necessary build.

If, as is apparently the case with polymer tanning agents, the tanning and filling effects can be integrated together with the softening effect into the substance, and not, as is currently the case in wet finishing, into the system, and should such a substance also exhibit hydrophobic properties, the conventional distinction between substance and process in tanning and fatliquoring could be consigned to the past.

The assessment of new developments in the area indicated must begin with consideration of the effects of drying and the mechanical processes of fiber isolation (staking, milling). The drying conditions have a considerable influence, positive or negative, upon the tendency, which is inherent to the collagen and retained to some extent by the leather matrix, to densify. The conflicting objectives of softness and yield will of necessity not be eliminated completely, but can perhaps be minimized.

The drying conditions which influence the structure, ultimately therefore "substance-related drying conditions" and/or the development of "substance, adapted to drying", constitute an area which to date has seen little in the way of in-depth study. The effects of the mechanical treatment processes, in particular staking and milling, are dependent upon the structure and consequently the substance, and require closer examination.

2.6 Interaction Equivalents of Active Substances with Collagen

2.6.1 The Special Case of Water

Water constitutes an integral part of the collagen structure, and as a softener and crosslinker, influences the later characteristics of the resulting leather. The *"bulk water"* acts as a solvent and is removed partly during the leather manufacture. The remaining water components are bound to the collagen with different degrees of stability. Initially, they block all sites capable of binding, albeit in a dynamic state of equilibrium. Substances capable of binding then displace this water in accordance with their bound quantity. A direct relationship should therefore exist between the fixation of the tanning material and the residual water content. Unfortunately, this relationship cannot be demonstrated reliably: a number of the tanning agents studied bind water for their own part, and only an overall balance can therefore be drawn up for the leather matrix.

Were all groups of collagen with the potential capacity to form hydrogen bonds (12.43 m_{Eq}/g) to bind one water molecule, the resulting water content would be 22.4% in terms of collagen. This is a plausible value for the hydrate water content of leather dried in a controlled climate in terms of its collagen content. It is a figure of little use, however: firstly, water does not attach itself in monolayers, but aggregated (in multilayers); secondly, by no means all groups potentially capable of forming a hydrogen bond are actually accessible. Calculations were therefore performed with regard to the interstitial deposition equivalents. If the base area of the water molecule is assumed to be 0.0225 nm^2, quantities of water in accordance with the various CSEs could be accommodated (Table 2.2).

Table 2.2 Water content of the CSE surface

Collagenous structural element, CSE	% Water, in terms of collagen of the CSE concerned
Molecule	292
Microfibril	108
Fibril	3.4
Elementary fiber	0.08
Fiber	0.002

2.6.2 Bond Equivalents Resulting from Ion Interaction

Low-molecular electrolyte of type HCOOH, H_2SO_4, $Ca(OH)_2$: all these compounds yield stoichiometric bond equivalents in principle. Selected comments on the underlying ion relationship follow.

In the case of a pure ion interaction, a direct relationship exists between the binding capacity of the collagen, expressed in m_{Eq}/g collagen, and the equivalent weight of the bound substances. The acid-binding capacity is 0.86 m_{Eq}/g collagen; the base-binding capacity is dependent upon the degree of deamination. As these are only fundamental model calculations, a bond equivalent of 0.85 m_{Eq}/g collagen was assumed for both. The diagram below shows the relationship between the equivalent weight of the compound concerned and the bound quantity per gram of collagen as a function of the collagenous functional groups involved (Figure 2.7).

Fundamental Collagen Chemistry in Leather Making

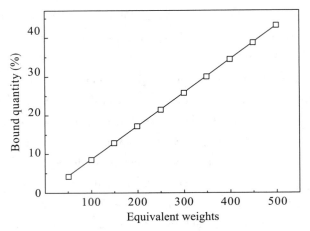

Figure 2.7 Relationship between equivalent weight and bound quantity, a bond equivalent of the collagen of 0.85 m_{Eq} served as an example

For low-molecular acids, such as sulphuric acid or formic acid, favorable quantities for application are in the order of magnitude of the bonding capacity of the collagen, as shown by the table below, which is based upon the supply in the pickle according to a typical recipe (full chrome auto leather). The supply results in virtually complete exploitation of the bonding capacity an acid reserve and consequently strong acidification of the pelt do not therefore arise (Table 2.3).

Table 2.3 Acid equivalents of the collagen in the pickle

Dimensions	Formic acid	Sulphuric acid
Equivalent weight	45	48
Bond equivalent on the collagen	3.8%	4.8%
Supply in terms of pelt mass	0.4% (at 100%)	1% (at 98%)
Supply in terms of collagen	1.3%	3.2%
Percentage exploitation of the acid-binding capacity of the collagen	34%	66%

The binding of higher-molecular ionic compounds would give rise to take-up quantities which are also fairly typical of practical conditions. This can be seen from the following example: a formaldehyde-condensed naphthalene sulphonic acid possesses an equivalent weight of 167 and 14.2% of the collagen could thus be bound. Should 4.25% be available to the pelt-well in line with industrial practice, this equates to 17% in terms of collagen, and therefore somewhat above the acid-binding capacity. This is, however, quite justifiable in the case of synthetic organic tanning materials.

The take-up of lime is to be considered in conjunction with the base-binding capacity. An enormous surplus of lime is available, owing to its limited solubility. The collagen is able to bind

Chapter 2 Physical and Chemical Processes on Collagen and Its Transformation into the Leather Matrix

approximately 1.7% at the equivalent weight of calcium of 20. This equates to approximately 0.5% in terms of pelt mass. Owing to the oversupply of lime, the float is always lime-saturated. The solubility of $Ca(OH)_2$ is 1.5 g/L, corresponding to 0.55 g of calcium per liter. At a water content of the pelt of 300% in terms of collagen, the ratio of collagen to water is 100 g to 300 cm^3 water with 0.16 g of calcium. This represents only 10% of the collagen's capacity to bind calcium, i. e. further calcium ions must be supplied from the (initially undissolved) lime surplus.

2.6.3 Bond Equivalents due to Covalent Bonding

HCOH, OHC—$(CH_2)_3$—CHO, ONC—$(CH_2)_4$—NCO: as an important modern tanning agent, glutaraldehyde was considered; formaldehyde was also considered, in particular owing to its small molecule size and the fact that it has been well studied. Di-isocyanates are of interest as "chemical compasses". These tanning materials also represent general examples of covalent bonding.

All covalently reacting substances for industrial use interact with the basic groups of the collagen (bond equivalent 0.85 m_{Eq}), preferably with the ε-amino groups of the collagen (Lys, Hyl; bond equivalent 0.33 m_{Eq}). At an M_w of the formaldehyde of 30 (= maximum equivalent weight and single-point binding, see above) or a minimum equivalent weight of 15 with two-point binding ("crosslinking"), quantities of between 0.49% are bound with crosslinking by Lys/Hyl and 2.6 with single-point binding by all amino groups of the collagen. With a supply—typical in industrial use—of 2% ~ 30% CH_2O in terms of pelt weight, equal to 2.4% pure formaldehyde in terms of collagen, this is an argument against crosslinking of "monomer" formaldehyde, even with the inclusion of all basic groups in the formaldehyde bonding. Conversely, "dimeric" formaldehyde would produce plausible binding quantities (W_{eq} = 30 with two-point binding) (Table 2.4).

Table 2.4 Binding of formaldehyde on collagen

Bond equivalents in collagen	Single-point or two-point binding of "dimeric" CH_2O (W_{eq} = 30)	Two-point binding (crosslinking) of "monomer" CH_2O (W_{eq} = 15)
All basic groups: 0.86 m_{Eq}	2.55%	1.29%
Lys, Hyl: 0.33 m_{Eq}	0.99%	0.49%

Calculations are provided below for selected variants involving tanning with glutaraldehyde (GTA) which assume different crosslinking and different homopolymerization of the GTA. The GTA supply common in industrial practice is seen to approximate to the theoretical quantity of a crosslinking monomeric glutaraldehyde (Table 2.5). As higher GTA supplies of up to 2% in terms of pelt weight (8% in terms of collagen) can still be bound without difficulty, the inclusion of all basic groups and/or autocondensation of the GTA are also a factor in the case of GTA. Attention is drawn in this context to the use of higher molecular aldehydes from starch,

Fundamental Collagen Chemistry in Leather Making

"glycogen and dextrin, and aldehyde protein condensates".

Table 2.5 Binding capacity of glutaraldehyde to collagen

Dimensions	Single-point binding	Two-point binding (crosslinking)
"Equivalent weight" of GTA	100	50
Equivalent weight (GTA)$_3$	300	150
GTA supply in terms of pelt	0.5% (at 100%)	150
GTA supply in terms of collagen	1.8% (at 100%)	150
Binding capacity for GTA (0.33 m_{Eq} Lys/Hyl)	3.3%	1.6%
Binding capacity for (GTA)$_3$ (0.33 m_{Eq} Lys/Hyl)	9.9%	4.8%
Binding capacity for GTA (0.86 m_{Eq} all basic amino acids)	8.6%	4.3%
Binding capacity for (GTA)$_3$ (0.86 m_{Eq} all basic amino acids)	25.8%	12.9%

2.6.4 Bond Equivalents based upon Metal Complex Bindings

Basic chromium sulphates of differing degrees of olation, $[(H_2O)_4-Cr-(OH)_2-Cr-(H_2O)_4]SO_4$ and $[(H_2O)_4-Cr-(OH)_2-Cr-(OH)_2-Cr-(OH)_2-Cr(H_2O)_4]SO_4$ will be considered; the current discussions regarding the actual size of the chromium complexes will not be addressed here. It is widely agreed that binucleate complexes play a major role in chrome tanning and possess a structure in accordance with Figure 2.8.

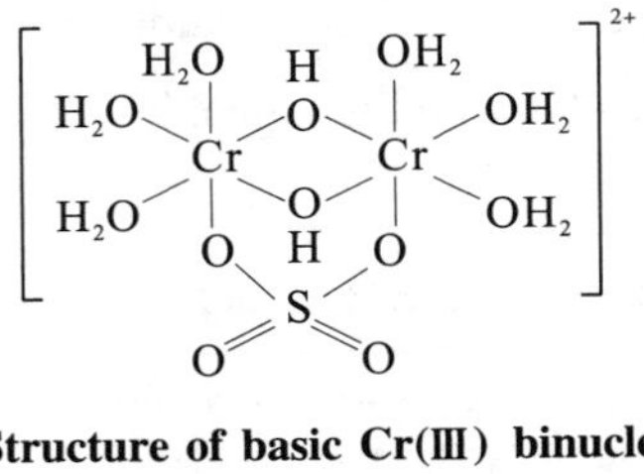

$$\begin{bmatrix} H_2O & H & OH_2 \\ H_2O & | & O & | & OH_2 \\ & Cr & & Cr & \\ H_2O & | & O & | & OH_2 \\ & O & H & O & \\ & & S & & \\ & O & & O & \end{bmatrix}^{2+}$$

Figure 2.8 Structure of basic Cr(III) binucleate complexes

These are the tanning agents of greatest practical significance. Their tanning chemistry has been closely studied[89], and their characteristics generally hold true for other mineral tanning materials which are bound to the collagen as metal complexes.

Owing to this particular practical and economic significance, the binding aspects of chrome tanning have been studied particularly closely: computer models have also been produced. The conclusion may be regarded as validated that the carboxyl groups of Glu and Asp, in particular, are involved in the binding of the chromium complexes, becoming their ligands. Their number is dependent upon the degree of deamidation of Gln and Asn in the lime. They also exhibit different pK values: their entry into the chromium complex is therefore dependent upon the pH. At this point, it is sufficient to assume a bond equivalent of 0.85 m_{Eq}/g collagen attributable to —COOH, and to consider the binding of binucleate and tetranucleate chromium complexes by single-point and two-point binding (crosslinking) and to relate them to typical industrial supplies of chrome tanning material (Cr_2O_3).

The bound quantity of chromium complexes was calculated as follows, based upon the common industrial data for the chromium supply in % Cr_2O_3 of the pelt weight:

- Cr_2O_3 ($M_w = 152$) corresponds to a binucleate complex as shown in Figure 2.8 ($M_w = 202$).
- The conversion factor is thus 1.33.
- A supply of 1.5% Cr_2O_3 in terms of pelt equates to 5.2% in terms of collagen.
- 5.2% Cr_2O_3 corresponds to 6.9% binucleate complex.

This relationship differs only slightly for tetranucleate complexes; the calculation will not therefore be repeated here. The situation is quite different in relation to the bonding capacity. Generally, it can be established that practical chrome tanning gives rise to chromium bonds (6.9%) which are plausible only for equivalences of two-point binding ("crosslinking") of binucleate complexes (8.6%). This confirms the dominant role of the binucleate complexes (Table 2.6). Corresponding calculations can be performed for other complex-bound mineral tanning agents. Common to all of them is that the leather-forming effect is produced by relatively small quantities which are well within the theoretical binding capacity of the collagen.

Table 2.6 Binding capacity of chromium tanning material to collagen

Dimension	Single-point binding	Two-point binding (crosslinking)
Equivalent weight Cr—Cr	202	101
Equivalent weight Cr—Cr—Cr—Cr	404	202
Binding capacity of the collagen for Cr—Cr	17%	8.6%
Binding capacity of the collagen for Cr—Cr—Cr—Cr	34%	17%
Supply under industrial conditions of 1.5% Cr_2O_3 in terms of pelt mass gives rise to chromium bonding to the collagen of	6.9%	17%

2.6.5 Bond Equivalents due to Combined Ionic and Hydrogen Bonding

Aromatic syntans, in particular those of the replacement tannin category, do contain functional —SO_3H groups which give rise to ionic bonds, as discussed above. However, they also contain phenolic hydroxyl groups which enable them to form hydrogen bonds.

A synthetic organic tanning material (replacement tannin, Figure 2.9) was therefore considered as an example substance from a major tanning material class with combined ionic and hydrogen bonding capacity.

Fundamental Collagen Chemistry in Leather Making

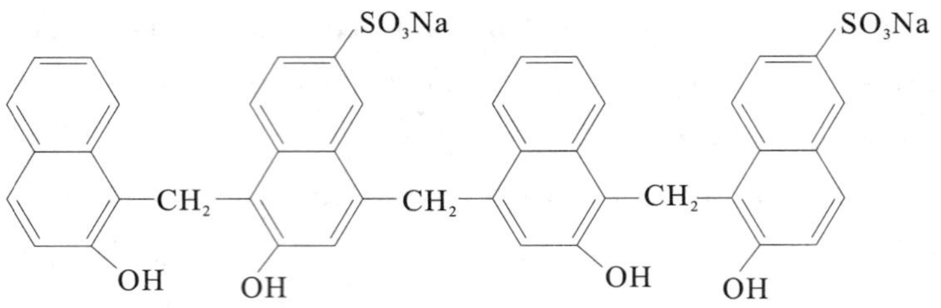

Figure 2.9 Synthetic replacement tannin

Theoretically, this synthetic organic tannin ($M_w = 814$) could be bound by one or two sulphonic acid groups and/or one to four hydroxyl groups. Binding could also be achieved by single-point, two-point or even multi-point binding. These different possibilities for binding give rise to differences in the quantities of bound replacement tannin (Table 2.7).

Table 2.7 Binding variants of replacement tannin (the maximum, purely theoretical OH m_{Eq} of the collagen of 12.4 was assumed; in brackets, the more realistic value of 0.4 is indicated)

Functional groups	W_{eq} with single-point binding	Binding to collagen in %	W_{eq} with two-point binding	Binding to collagen in %
1 ×—SO$_3$H	814	70	–	–
2 ×—SO$_3$H	407	35	407	35
1 ×—OH	814	1009 (32)	–	–
2 ×—OH	407	505 (16)	202	250 (9)
3 ×—OH	271	336 (13)	135	167 (5)
4 ×—OH	204	253 (10)	102	126 (4)

Quantities used industrially range between 10% replacement tannin in terms of shaved weight (22% in terms of collagen) during retannage and 30% in terms of pelt weight (120% in terms of collagen) during heavy leather tannage. This means that binding by hydrogen bonds is always the binding variable which determines the quantity. The question remains open however as to how many of the collagen groups rendered capable of forming hydrogen bonds (Table 1.5), and how many on the synthetic organic tannin material side, actually enter into a bond. The calculation of bond equivalents is therefore questionable. These difficulties are even greater where vegetable tanning materials are concerned. A solution is anticipated involving the calculation of attachment and interstitial deposition equivalents; this will however probably be attainable at some stage, but only by means of computer simulation.

Owing to the broad similarity to synthetic organic tannin materials, dyestuff binding will not be addressed in any greater detail. Three points should be noted, however: the dye shown in Figure 2.10 is, like many others, essentially replacement tannin with good tanning capacity. The above remarks thus apply in full to synthetic organic tanning materials. The notable point here is, however, that it possesses more —SO_3 groups and fewer —OH groups than the synthetic organic tannin material, yet still has a good tanning action. This underlines the influence of the dipole character of the aromatic tanning materials upon tanning material fixation, i. e. the role of the spatial arrangement of the tanning materials bound by secondary valences. This is an area which deserves further study.

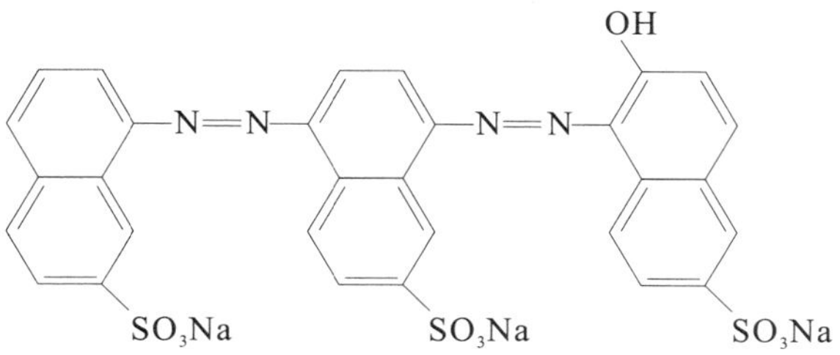

Figure 2.10 Diamino gene blue dye as an example of a dye also active as a tanning material with several functional groups

Finally, attention is drawn to the fact that dyes of this constitution also bind as ligands to the chromium complexes in the leather matrix. If an equivalent weight of the dye of 740 is assumed (binding by an —SO_3H group), the following could be bound in the leather to the chromium complex in accordance with the following calculation:

• Chromium content in leather 6.0% Cr_2O_3, collagen content 62%, Cr_2O_3 in terms of collagen = 9.7%.

• 9.7% Cr_2O_3 equates to 12.6% chromium complex.

• 12.6% chromium complex represents 0.06 bond equivalents (12.6/202).

• At an equivalent weight of the dye of 714, up to 43% of the dye stuff in terms of collagen would be bound.

• In practice, a maximum of 10% may be assumed.

• Determining factors include the low supply (ultimately, the process is that of dyeing and not of tanning; what is the limit of the take-up capacity?) and multifunctional binding, i. e. correspondingly low equivalent weight.

2.6.6 Bond Equivalents of Secondary Valence Bonds, in Particular Hydrogen Bonding

Two representative examples of the important tanning material class of vegetable tanning materials are considered here: quebracho catechin, being a relatively low-molecular compound, and castalagine as a high-molecular compound.

The calculation of equivalences in this context poses three problems. Firstly, the type of secondary valence forces arising during interaction as a function of the substances is not clearly defined. Although the dominant role of hydrogen bond formation in the binding of phenolic tanning materials has now largely been validated, relatively minor differences in the dipole character of the individual aromatic substances (dyestuffs, tanning materials) appear to influence their binding and therefore their tanning capacity. Likewise, the quantitative significance of the van der Waals' and hydrophobic action has also not been adequately clarified and may require further study. For the reasons stated above, studies of equivalences were not performed here, particularly as they have been dealt with by way of an example for replacement tannin.

The second problem also affects the relatively clear case of pure hydrogen bonding and concerns the number and availability of functional groups of the collagen, i. e. in this case capable of hydrogen bond formation, as already discussed above.

Finally, the question remains regarding the number of phenolic hydroxyl groups of the tanning material to be considered for calculation of the bond equivalents. Only with more exact knowledge of the spatial conditions in the tanning material and on the collagen (in this case, sequence distribution on the individual structural elements) can more detailed conclusions be expected from computer modeling.

Polymer tanning materials are growing in practical importance; those available differ widely in their chemistry and structure. They are generally regarded as binding chemically by their carboxyl groups to the free ligand sites in the chromium complex (exchange of aquo groups). This can be estimated by analysis, as shown by the example of dyes; identification of the equivalent weights is essential in this case. If a figure of 300 is assumed, the binding capacity would be in the order of 18 according to the model adopted above for dye binding, which is quite plausible. Other results are supplied by the attachment and interstitial deposition equivalents presented below, which will be addressed later.

For surfactants, fats, and indifferent fillers such as sulphur and kaolin, calculation of (chemical) bond equivalents are not worthwhile. Their interaction with the collagen is taken into account much more effectively by the attachment and interstitial deposition equivalents considered below.

2.6.7 Attachment Equivalents: Possibilities for Bonding on Collagen Surfaces

Parallel and/or in addition to binding in the context of chemical equivalences, attachments may occur on the collagen surface of the individual collagenous structural elements. This applies in particular to the substances described, which are bound by secondary valences. Such deposits limited to the surface are however also described for example for basic chromium salts. In order to provide a preliminary survey of the quantities which can be bound in this case, the calculations discussed at the outset were performed for selected low, medium and high-molecular compounds (Table 2.8).

Chapter 2 Physical and Chemical Processes on Collagen and Its Transformation into the Leather Matrix

Table 2.8 Attachment equivalents of various substances on the different CSEs as a percentage of the collagen of the individual CSEs

Substance/CSE	Molecule	Microfibril	Fibril	Elementary fiber	Fiber
N_2	236	87	3	0.07	0.006
H_2O	292	108	3	0.08	0.002
Chrome tanning material	183	67	2	0.05	0.001
Replacement tannin	132	49	1	0.04	0.0009
Quebracho catechin	31	11	0.4	0.009	0.0002
Castalagine	48	18	0.6	0.01	0.0003
Polymer tanning material I	1380	507	16	0.4	0.01
Polymer tanning material II	891	328	10	0.2	0.006
C_{18} fatty alcohol sulphate	60	22	0.7	0.02	0.0004
C_{18} triglyceride	578	213	7	0.2	0.004

The example of water has already been alluded to with regard to the fundamental relationship between the (specific) surface area of the individual CSEs and their attachment capacity. A regular dependence also exists regarding the molecule size (M_s) and the relative molar mass (M_w), and also regarding the orientation of "box-like" substances (parallel and perpendicular orientation). Where certain differences exist between the individual substances analyzed, the quantities of the substances to be attached to the fibers and elementary fibers are in all cases low, and by no means typical of industrial application. The opposite is true of their capacity to attach to molecules: this is too high in all cases, even without consideration of problems of intermolecular space. The conditions for microfibrils and fibrils appear more realistic in terms of their greater similarity to industrial conditions.

Similar observations have already been made for a C_{18} fatty alcohol sulphate and for alkyl succinate, and practical conclusions drawn for fatliquoring; the conclusion in this case was that far too much fat is used if only a superficial lubricating effect in the fibrillar region is to be attained.

The data available are certainly very illustrative, and warrant discussion with regard to the substances, e. g. with regard to the relationship between morphology and relative molar mass in the case of polymers, or the validity of the assumptions for kaolin. Care should however be taken not to over-interpret the relationships in tanning theory. This aspect must also be considered in conjunction with the results, described below, of the calculations of interstitial deposition equivalents.

2.6.8 Interstitial Deposition Equivalents: Facility for Attachment in the Various Voids of the Collagen

It should in principle be possible to "fill" the pores and capillaries present in the individual structural levels. The possibilities created by this mechanism for the binding of substance ("interstitial deposition equivalents") naturally exceed the attachment equivalents shown above; multilayer coverage represents a smooth transition to pure filling of the voids. This holds true, of course, only when the substance dimensions do not exceed those of the void concerned. Selected example calculations are shown which employ the pore distribution. The pore volume concerned was first divided here by the molecular dimension of the substance concerned, the resulting number of molecules multiplied by the molecular mass, and the quantity thus obtained expressed as a% of collagen (Table 2.9).

Table 2.9 Interstitial deposition equivalents in% in terms of collagen of different substances in various void categories, in the latter case

Substance/Pore category	2.5~5.0 μm 0.44 cm³/g (31%)	>0.1 μm 0.12 cm³/g (9%)	Total porosity 1.4 cm³/g (100%)	0.25~0.5 μm 0.16 cm³/g (11%)
Water	387%	105%	1230%	141%
Castalane	3.8%	1%	12%	1.4%
Polymer tanning material I	179%	48%	477%	65%
Polymer tanning material II	28%	7.6%	89%	10%
C_{18} triglyceride	41%	11%	131%	15%

The results are nor particularly convincing. For instance, the water content obtained in this way is purely arithmetically too high since, as can be easily demonstrated (Table 2.10), it may only be 140% for the total porosity. In the case of castalane, triglyceride and polymer tanning material II, the results appear "reasonable", whilst still differing substantially from those produced by the much simpler method described below. The reasons should be made the subject of a joint discussion; they may have their origin in unsatisfactory data for the porosity or for the substances, or in the mathematical method. As described in the introduction, the second method is based upon multiplication of the void volume by the substance density. This was demonstrated by way of example for the overall porosity. The values obtained by this method are plausible, if it is considered that they relate to a total fill, which would not be encountered in practice.

Chapter 2 Physical and Chemical Processes on Collagen and Its Transformation into the Leather Matrix

Table 2.10 Interstitial deposition equivalents as a function of the void dimensions and substance density

Substances/Data	Void per gramme of collagen in cm^3	Substance density in g/cm^3	Substance mass at total fill in g	%Interstitial deposition, in terms of collagen
Water	1.4	1.0	1.4	140
Castalane	1.4	1.6	2.2	220
Polymer tanning material I	1.4	1.3	1.8	180
Polymer tanning material II	1.4	1.3	1.8	180
Triglyceride	1.4	0.9	1.3	130

Attention is drawn to the fact that initially, all processes occur in the fully hydrated collagen with maximum distances between the CSEs, but that these distances may be modified by swelling or depletion. This phenomenon is the source of considerable uncertainty. The planned theoretical studies into the part played by water in the tanning processes should enable this uncertainty to be reduced to a minimum.

2.6.9 Penetration Considerations

The conditions which must be met before a substance can even reach its location of activity within the collagen structure (see "diffusion funnel" in Figure 2.6), referred to in the main text, are summarized again here for the sake of clarity. The dimensions of the substances considered on the present data sheet, have been summarized and compared to the void categories in accordance with Table 2.1. The majority of industrially significant tanning materials are seen to be able to penetrate the intrafibrillar cavities, but compounds are also shown to retain potential tanning action when they act on the interfibrillar level of the collagen structure. Substances whose action is predominately that of a filler are in any case deposited between the fibrils, and also between the elementary fibers. Conversely, the interfiber cavities remain largely unfilled and give rise to the porosity of the leather, its chief characteristic. Attention is drawn again in this context to the concept, elaborated in the introduction, of the interaction with N_2 during assessment of the specific surface area by means of BET measurement (Table 2.11).

Table 2.11 Possibilities for collagen penetration as a function of the particle and pore size

Substance category	Mean particle size (examples)	Nanopores (up to 1.7 nm)	Micropores (up to 100 nm)	Pores (1 ~ 3 μm)	Macropores (>3 μm)
Nanoparticles	<0.15 nm, N_2, H_2O, CH_2O	Intramolecular conversion possible and observed or postulated			
Microparticles I	0.15 ~ 17 nm, GTA, chromium, etc.	Intrafibrillar conversion possible and demonstrated in numerous cases			
Microparticles II	17 ~ 100 nm, all phenolic tanning agents, where not aggregated, fats, polymer tanning materials		Interfibrillar conversion quite frequent, transition to filling blurred		
Macroparticles	>100 nm, all self-aggregating (phenolic) tanning agents, kaolin, and also pure fillers			Encountered as filler with certain leather types	"Real" unfilled cavities; as porosity of the leather; its chief characteristic

Attention is drawn to another two aspects. The first of these is an aggregation of numerous phenolic tanning materials and dyes. The particle sizes actually present within the diffusion paths of the pelt have not been ascertained satisfactorily. Conversely, the considerations and calculations presented here all demonstrate that the processes arising on the collagen considered by the present study may take place in quite different regions of the collagen structure. Polydisperse systems are seen to be particularly well suited to promoting an optimum tanning action in the interests of structural stabilization of the collagen.

2.7 Summary

The transformation of collagen into the leather matrix was considered with regard to chemical and structural aspects. Certain basic relationships regarding substances penetrating into the interior of the collagen were discussed. Based upon dimensions of the individual structural elements of the collagen which were known and/or calculated for the purpose, the interstices between them, and data on the active substances, the possibilities for bonding were determined from chemical and steric perspectives. Data sheets of the resulting data in tabular form were produced, thus providing the facility in the future for rapid assessment of the possibilities and scale of material, structurally relevant interactions with the collagen, and its transformation into the leather matrix.

Chapter 3　The Theory of Tanning
—Past, Present, Future

3.1　Introduction

"A clear and proper understanding of the tanning process must influence motivation and the direction of technical progress in the leather industry."

—E. Stiasny

When Stiasny made the above statement in 1935, one which would still be met with universal agreement today, he expressed at the same time his regret that the "question regarding the exact nature of the tanning process has still not been satisfactorily answered". Whether this is still the case today was the motivation for production of the present progress report.

Tanning constitutes only one stage in the transformation of collagen into leather, but it remains an important one. It is not surprising that for a long time, "tanning" was virtually synonymous with "leather production", as the tanner and leathermaker were one and the same. Vegetable-tanned leathers, which dominated for thousands of years, owed their characteristics to the tanning material first and foremost owing to its presence in high quantities in the leather, for example, in a ratio of almost 1:1 to the hide substance in the case of sole leather. The characteristics of harness leather and other vegetable-tanned leathers were determined overwhelmingly by the tanning process. The characteristics of glace leather, leather tanned with alum, in some respects a forerunner to modern compact retannage and fatliquoring procedures, were generally regarded as being determined by the tanning process. The same applied to the characteristics of chamois leathers, which were simultaneously tanned and softened by the fish oils. It should therefore come as no surprise that the decisive role of tanning has been emphasized in more recent times not least in chrome leathers, as chrome tanning endowed the leathers with hitherto unknown characteristics. That this form of tannage owed its breakthrough in part to the development of fatliquoring had little influence upon its perceived importance.

Given this function of tannage, whether actual or presumed, it comes as no surprise that with the rise of modern science, tanning researchers repeatedly endeavored to throw some light on the theoretical principles of this process with its decisive significance for leather manufacture. Their efforts always began with the description of the observed effects of the

Fundamental Collagen Chemistry in Leather Making

interaction between skin collagen and leather; tanning theories were and remain defined primarily by phenomena. Their physical and chemical underpinnings were thus always dependent upon knowledge of the tanning materials concerned, which grew considerably in number over the last hundred years or more, only few of which, however, became or remained of practical significance (Table 3.1). However, these underpinnings were primarily dependent upon knowledge of collagen. They were promoted by the constant, ongoing development of analytical research methods in general and of tanning chemistry research methods in particular.

Table 3.1 The most important tanning materials: The date of their first practical or proposed use, their current significant and the current knowledge of tanning theory

Tanning materials (tanning processes)	Since when proposed or in practical use	Practical significance today	Level of knowledge of tanning theory
Vegetable tanning materials (bark tannage)	Known since time immemorial	Major, particularly in retannage	Hydrogen bond formation dominates
Aluminum salts (tawing)	Known since time immemorial	Dyeing auxiliary, buffing quality enhancer, fur finishing	Complex bonding, less fast than Cr
Fish oils (chamois tannage)	Known since time immemorial	Constantly low (niche products)	Aldehyde tanning predominates, details still not clarified
Iron (III) salts (iron tannage)	Johnson 1770, Knapp 1858, Ferrigan 1940	Constantly low (niche products)	Can be used highly masked only
Formaldehyde	Payne and Puttmann 1898	None as sole tanning agent, low as cleavage product	Covalent bonding
Cr(III) salts (chrome tannage)	Knapp 1858, industrially significant since 1885	Dominant tanning process	Complex bonding; comprehensively studied, but with questions as yet unanswered
Quinones	Meunier and Seyewitz, 1908	No direct significance, in some cases contributory in vegetable tannage	Covalent bonding
Sulphur tannage	Eitner 1911, sporadic earlier cases	Virtually none	Unanswered questions remain
Silicic acid tanning	Hough 1919 (earlier references: Graham 1862)	Virtually none (to date)	Unanswered questions remain
Aromatic syntans	Stiasny 1913	Major, particularly in retannage	Salt and hydrogen bonding
Zirconium salts	Bayer 1935	Minor, in special cases only	Complex formation, details open

50

Chapter 3 The Theory of Tanning—Past, Present, Future

Continued

Tanning materials (tanning processes)	Since when proposed or in practical use	Practical significance today	Level of knowledge of tanning theory
Polymer tanning agents	DuPont 1938, acrylic syntans (lubricating tanning agents) since 1975	Major, particularly in retannage	Unanswered questions remain
Polyphosphates	1938 (Germany, USA)	Relatively minor	Salt binding
Resin tanning materials	Since 1940 (USA)	Considerable	Involved in CH_2O effect
Paraffin sulphochloride (Immergarn process)	Immendörfer (BASF) 1943	None as tanning agent, in some cases as fatliquoring agent	In some case covalent bonding
Titanium salts	1950 (USSR)	Regional only, minor	Complex formation, details open
Diisocyanates, oligomer and capped polyurethanes	Bayer 1944, Bayer since 1960	Virtually none, very minor	Covalent bonding
Glutaraldehyde (including diadehydes)	Filachione 1960	Considerable (FOC)	Covalent bonding
Oxazolidines	1980	None (to date)	Mutual action with vegetable tanning agent, details unclear, involved in CH_2O effect
THP salts	1985	Virtually none (to date)	Covalent bonding, involved in CH_2O effect

The first publication to deal with tanning theory is one by Seguin and dates back to 1795; in 2001, Ramasami dedicated the 48th J. A. Wilson Memorial Lecture to the establishment of a "Unified Theory of Tanning"[90]. The most important papers published during the good two centuries between these dates are the work of familiar scientists. Names such as Knapp, Procter, Stiasny, Wilson, Küntzel, Gustavson, Grassmann, Stather, Heidemann and Covington are notable examples. They reflect the progress of scientific knowledge of their time. The early papers are among those of the most meticulous of observers and shrewdest of thinkers; a study of them thus provides insights which remain valid today and, in some cases, have unjustifiably been forgotten.

The objective of the present progress report is to summarize briefly the development of tanning theory in terms of results that have survived the test of time, to describe its current status critically in terms of its practical consequences, and to set out in brief future tasks.

3.2 The History of Tanning Theory

The astringency of oldest tanning agents, alum and vegetable extracts, was known in antiquity. The famous physician Galenus concluded in the second century A. D. that the astringent effect of these substances was responsible for the tanning action. Not until 1795, however, was a more exact study published by Seguin, a tanner with training in chemistry, being as he was a student of Lavoisier, whose accomplishments included introducing scales to the chemical laboratory. Seguin was the first to propose, correctly, the existence of an analogous reaction between the tanning of hide and the precipitation of tannin by means of gelatine. His explanation of this reaction as the formation of salt between the basic hide and the acid tanning agent is attributable to the wisdom of his time, but was developed another 100 years later in the form of the "Procter-Wilson tanning theory". Of greater importance is his second discovery, namely that tanning is always associated with an increase in the weight of the dry hide (pelt):

$$mass\ (pelt\text{-}water + tanning\ material) > mass\ (pelt\text{-}water)$$

Trivial, but ignored up until that point and sometimes even now.

It was Friedrich Knapp who then in 1858 made a lasting contribution on the subject of the nature and substance of tanning and of leather, as reflected in the title of his famous paper. To Knapp, we owe the emphasis upon two essential effects of tanning: firstly, it prevents the hide fibers from adhering to each other during drying, and thus guarantees, at the latest following mechanical staking, the soft, supple and thereafter opaque drying-out of tanned pelt in comparison to the hard, stiff, transparent property of dried untanned pelt.

At the same time, tanning eliminates more or less completely the tendency of wet hide to putrefy. Knapp also points out that tanning is not necessarily required in order to bring about leatherlike drying-out: a similar effect can be achieved by the use of alcohol or concentrated salt solutions. Knapp attained the water resistance of leatherlike drying-out by dehydrating with alcohol in the presence of stearic acid: an idea applied in contemporary form 100 years later by Heidemann and Riess with silicon treatment of pelt dehydrated with acetone. In fact, 1% stearic acid absorbed by the pelt was sufficient for attainment of a bright white, dressed, fine-grain leather. Knapp concluded correctly that a universal chemical theory of tanning could not exist in view of the large number of chemically diverse tanning agents. He discovered the tanning action of the basic chromium and iron salts. And that in fact a chemical reaction was not an absolute requirement for leatherlike drying-out. He regarded the latter case as applicable when the collagen fibers are coated in a waterproof manner (the "coating" theory).

The "chemical tanning theory" thus established by Seguin with his concept of salt formation as the essence of tanning and Knapp's "physical tanning theory", in each case modified by the results of studies on new tanning agents and tanning processes (metal salts, aldehydes, quinones), subsequently shaped the discussion of tanning theory over the following decades. The tone in the relevant publications remained passionate. In 1918, Moeller subjected his fellow-researcher Griffith, who declared chamois tanning to be a form of aldehyde tanning: "this theory, promoted in the absence of any experimental basis, continues to haunt the minds of theorists, and Griffith persists in preaching the doctrine that aldehyde tanning, as a mutation of chamois tanning, is among the oldest of tanning processes." Today, we know that he was right.

Chapter 3 The Theory of Tanning—Past, Present, Future

The lack of detailed knowledge of collagen and inadequate knowledge of the active tanning materials led to many an absurd theory. On the basis of the apparent interaction between collagen and quinone, for instance, Fahrion, Meunier and others explained tanning in general as collagen oxidation. Conversely, Fahrion likewise demonstrated that only the breakup of the hide's structure brought about by lime and bating liberated the reactive groups required for chemical interaction with the tanning agents.

It was also an era in which chrome tanning began its industrial and economic conquest and attracted corresponding attention from the scientific community. The detailed studies of aldehyde tanning performed by Gerngross, notable for the methods employed, and those of quinone tanning performed by Meunier and Seyewetz were also a product of this period. In all three cases, highly relevant conclusions were drawn concerning the underlying reaction mechanisms (complex formation in the case of chrome, covalent bonding in that of aldehydes and quinones).

The demonstration that the adsorption curves of tanning agent by (pelt) collagen corresponded extremely closely to those of other adsorption processes appeared to be a neat confirmation of the physical coating theory: it seemed logical after all to apply the physical and chemical findings of Gibbs, Haber and Langmuir in particular, concerning the nature of adsorption processes and the interaction between adsorbate and adsorbent, to tanning processes.

Since tanning, in contrast to many other adsorption processes, is, however, irreversible, the tanning agent must be bound chemically to the collagen following its adsorption, and/or undergo a further, secondary change. Discussions of colloid chemistry aspects thus became more important: the vegetable tanning agents, the (olated) chromium (metal) complexes and also fatliquors with tanning action are, after all, colloid-disperse substances which are deposited largely in insoluble form following primary adsorption. This interpretation of the role of the colloid dimension as a precondition for the tanning capacity was supported, for example, by precipitation tests on gelatine, solutions: tanning has a precipitation effect, gallic acid does not; although they contain comparable groups, they differ in their particle size. Silicic acid precipitates gelatine only once the colloid form has been produced by heating ageing. Work conducted by Stiasny in particular brought the role of the constitution and the particle size (the "degree of distribution") into the discussion, and the boundaries between the physical (coating) theory and the chemical theory were blurred and indeed rendered meaningless.

Küntzel was then able to demonstrate by studies of the fiber birefringence that the collagen-tanning agent interaction was not restricted to the fiber surface, as originally postulated by the adsorption theory, but that a reaction ("permutoid" according to Freundlich) takes place as a function of the particle size of the tanning materials, through to the innermost regions of the collagen micro-structure, about which little was known at that time.

In the mid-1930s, the view became established that a secondary or primary valence interaction between tanning materials and collagen constituted the essential characteristic of tanning. Stiasny, like Knapp before him, drew from this the conclusion which still stands:

"A theory which satisfies all mechanisms for conversion of hide into leather is seen not to exist, and it will be necessary to assign the various modes of tanning to different understandings of the tanning processes."

It is also interesting to note how progress in chemical knowledge is also reflected in successive developments in the discussion of tanning theory. A particularly illustrative example is a series of publications by Wilson, in which he attempted to apply the rapid developments in electron theory (Lewis-Langmuir) of his time to tanning theory. By accident, he departed from his former view of salt formation to that of hydrogen bond formation as the mechanism by which vegetable tanning agent is bound to the collagen.

The following 20 years were a period of consolidation of tanning theories, generally in relation to specific tanning processes. This progress was promoted by the rapid growth in knowledge of collagen and by new analytical techniques. Küntzel for example, in a series of papers on the nature of chrome tanning, also made a substantial contribution to tanning theory. Closer attention will therefore be paid here to certain important results.

The helix-coil transition brought about by the melting of gelatine gels (gel-sol transition at a temperature transition from $T < 15$ ℃ to $T > 35$ ℃ is associated with a change in the specific rotation $[\alpha D]$. The quotient of $[\alpha D15]$ and $[\alpha D35]$ is termed mutarotation and reflects the conformational change. Küntzel demonstrated that this mutarotation is hardly impaired even in a gelatine rendered boilfast by chrome tanning. This means simply that even small quantities of chromium have a substantial influence upon the hydrothermal stability of gelatines even though the greater part of the gelatines is not affected by chrome tannage. Küntzel applied this observation to the fibrous pelt collagen: low quantities of chromium give rise to high T_s, large regions of the collagen are left untanned. The fact that Küntzel assumed "laterally untanned" interior regions, where we now think in terms of "longitudinal cluster, is attributable to the incomplete knowledge of collagen of his time, which incidentally he explained expertly himself and linked to concepts of tanning theory".

Küntzel's contributions were also crucial in leading to an understanding of the nature of leather like drying-out of tanned skin collagen. His paper on the cause of leather like drying-out of chrome-tanned hide remains to this day one of the ground-breaking publications on tanning theory. He provided the following plausible explanation for the fact, already known, that the fibers of the tanned pelt, in contrast to those of untanned pelt, do not adhere to each other during drying-out: "It must first be stated that complete adherence of the fibers to each other during drying-out of the pelt is due to the surface tension of the water on the one hand, and to the softness and suppleness of the untanned fibers on the other: in the same way as the hairs of a fine brush spread apart in water but close up when the brush is removed from it owning to the water's surface tension, the exceptionally soft and yielding fibers of the pelt close up in the wet state when the water is removed from the voids between the fibers owing to evaporation to the atmosphere."

He then refers to the stiffening of the "mesh of hide substance", or fibrils as we would now describe them, recognizable by individual fibers which behave like "thin rigid wires" following tanning, and to the associated fixing of the form (e. g. "tanning wrinkles"). He then continues: "Owing to this stiffening brought about by the tanning materials, the hide fibers are able to resist the deformative force of the surface tension of the water; consequently, the fibers do not adhere to each other. The fiber network dries out in a leatherlike manner. An illustrative example of this kind of drying-out is that of a brush with stiff bristles which remain separate when the brush is withdrawn from water owing to their stiffness."

The effect of dehydration with solvent or the action of acid/salt as in the Leipzig pickling process, which also has the effect of fiber isolation, is attributed by Küntzel to the fact that during dehydration the fibrils shrink concentrically and are hardened in the process, a fact which has long been known from microscopic study. In contrast to the water-resistant effect of tanning materials, the effect of treatment with solvent and pickle finishing is fully reversible (hence "pseudo-tanning"), unless the leatherlike drying-out is rendered "permanent" by simultaneous or subsequent hydrophobing, as already mentioned above.

After presenting comprehensive discussions of tanning theory and the nature of leather formation in the Bergmann-Grassmann manual, Gustavson compiled two volumes in the late 1950s containing the knowledge of his time on collagen and the nature of the tanning process[91, 92]. Attention is also drawn to A. N. Michailow's excellent book. This work covers the poorly accessible Russian literature and is notable for the whole series of original approaches which it contains.

Knowledge of collagen continued to develop rapidly into the early 1970s. The discovery of soluble collagen and thus the existence of a discrete collagen molecule, the complete determination of the amino acid composition and sequence, and explanation thanks to electron microscopy of the mechanism of fibril formation also represented milestones in tanning chemistry, and were decisively fruitful additions to the knowledge of tanning and leather formation.

Collagen research then proceeded on an undreamed-of scale once collagen had been identified as the biologically significant component of the extracellular matrix. Collagen ceased to be a single substance and became a family of some 20 members in the mammal branch alone. The majority of findings, however, ceased to be relevant for tanning chemistry and the "Type I" skin collagen relevant to it. In other words, the knowledge of collagen required for development of a theory of tanning has essentially been available since the 1970s (Table 3.2) [93, 94].

Fundamental Collagen Chemistry in Leather Making

Table 3.2 Milestones in collagen research and their relevance to the development of tanning theory[1]

Time frame	State of and growth in knowledge	Consequences for tanning theory
Up to around 1900	Primarily histological knowledge	Collagen is a fiber network; fiber isolation as the characteristic of tanning
Up to around 1940	Increase in chemical knowledge: amino acids are linked to form peptides; the "micelles" can also be found below the fiber level	Tanning is also a chemical process: reactions on amino acids and peptides serve as tanning models
Up to around 1960	Discovery of the soluble collagens and thus of the discrete collagen molecule; elucidation of the complete amino acid composition; TEM permits insights into the microstructure and mechanism of fibril formation; developing the triple helix concept	Elucidation of the binding mechanism of the majority of tanning agents; findings concerning the processes in the individual structural levels of the collagen
Up to around 1970	Consolidation of knowledge, elucidation of the amino acid sequences, structural details of the collagen molecule and the fibrils; scanning electron micrography as a useful "magnifying glass"	General consolidation of tanning theory conceptions now possible. Essential basis of modern conceptions developed
Since 1970	Collagen increasingly recognized as an important component of the extracellular matrix. Consequent expansion of microbiological collagen research: collagen types and "family". Further detailed discoveries concerning structure, sequence and genesis	Refinement of conceptions of tanning theory; recognition of the role of water as an integral constituent of collagen, including for tanning. Exploitation of modern analysis procedures, deduced in some cases from collagen research
Present and future	Collagen research remains dynamic; increasingly oriented owing to medical materials-oriented wing to medical technology, and thus relevant once more to leather	Observance of developments, following up of analysis in particular (assembly, AFM, stress measurement, etc.) and observance of protein-water interactions

An original tanning theory and one the subject of heated discussion in its day is that of Batzer, a plastics chemist and student of Staudinger in Freiburg. He became known to the leather industry with his "polymer tanning". This process consisted essentially of monomers, such as methacrylate, being brought into pelt and subsequently being polymerized *in-situ* in the presence of redox catalysts. Given suitable monomers and reaction conditions, leathers of more than adequate softness were obtained, or the leather characteristics significantly improved. His conceptions, which were oriented towards polymer chemistry, prompted him to tanning theory considerations. He defined tanning for example as a form of "dehydrating softening". His argument was as follows: pelt contains a great quantity of water and is soft and flexible; water is therefore a softener. Leather contains no water and is also to a greater or lesser degree soft. Tanning agent has thus assumed the softening function of the water. It cannot be bound by crosslinking, as it would otherwise have given rise to hardening, a fact inferred by Batzer from examples in plastics. In a comment on this paper, Küntzel draws attention to a fundamental error on Batzer's part: pelt collagen is not strongly dehydrated by tanning. Rather, the crucial removal of water occurs during the drying-out stage. The fact that, in contrast to pelt, the tanned substrate remains to a greater or lesser degree soft in the process, is attributable to the reasons stated above. Küntzel then astutely goes on to disprove the theory of noncrosslinking of the collagen by tanning agents by drawing attention to the completely different conditions for proteins and polymers (rubber serving as an example). Of further note is that tanning does in fact give rise to a certain tangible reinforcement of the structure, which is in fact a condition for partial mechanical dehydration by dewatering and for subsequent shaving. Küntzel's reply is of importance to tanning theory and is worth reading even today. Batzer's theory, by contrast, has disappeared into obscurity.

Grassmann, who himself together with his Munich school—mention is made here of Kühn, Nordwig, Hannig, Hörmann, Endres, and others—at the Max-Planck-Institut für Eiweiss-und Lederforschung made a contribution of his own to the expansion of knowledge of collagen, presented the knowledge of the time concerning the nature of tanning in a notable lecture in 1960 under the heading "Modern theories of the mechanism of the tanning process". Based upon the arguments already mentioned, of his predecessors and his own work, he formulated in agreement with the other tanning researchers of his time a number of postulates which for the most part still hold true. He first compared the characteristics of skin and leather, and set out as characteristics of the formation of leather, upon which he commented the reduced swelling capacity, the leatherlike drying-out, the resistance to fermentation and putrefaction, and the rise in the T_s to >62 °C.

In Grassman's view, the tanning materials bring about these changes in the characteristics of the collagen chiefly by three mechanisms: blocking of hydrophile anionic or cationic groups of the collagen and thus hydrophobing of the collagen, stabilizing of the tertiary structure of the collagen (fibrils, elementary fibers, fibers) by the introduction of crosslinks, and coating of the fibrils, the last of these in some cases without direct interaction between tanning material and collagen ("pseudo tanning").

Finally, as far as the chemical nature of the tanning materials and their binding to the collagen is concerned, Grassman distinguishes between the substances which react covalently, the polymolecular, generally multivalent ions which react electrostatically, the metal cations capable of complex bonding, and the phenolic-aromatic hydrogen bond formers. This principle of classifying the tanning materials according to the nature of their bond to the collagen remains valid to this day. Based upon it, he describes a great many details of experimental verification of the assumed reaction processes and binding mechanisms.

At approximately the same time, a paper was published by Stather and Pauligk which aroused great interest and addressed the question regarding the minimum tanning material quantities required for conversion of hide to leather. Tanning experiments on hide meal and native hide material containing varying components of chromium salts, formaldehyde and various synthetic and natural phenolic tanning materials were assessed against the chief criteria of leather formation. These criteria were: leatherlike drying-out, characterized by measurement of the transparency and the apparent density as a measure of the retained porosity; stability under the influence of hydrolytic and fermentative attack by measurement of the resistance to hot water, alkalis and enzymes (papain), and the T_s. To conclude, characteristic changes to the collagen, described not entirely unproblematically by the authors as leather formation proper, can be attained with far lower quantities of tanning material than those required for production of a serviceable standard commercial leather (Table 3.3).

Table 3.3 Minimum tanning material quantities required for leather formation[1]

Tanning material type	Minimum tanning material quantity, in terms of hide substance
Basic chromium sulphate	0.9% ~ 1.25% Cr_2O_3
Formaldehyde	0.7% ~ 0.8% CH_2O
Various syntans and vegetable tanning agents	Dependent upon type: 6% ~ 12%, in all cases < 20%

To sum up, it is emphasized that chemical and physical changes in the collagen occur in parallel and that where the quantity of available tanning material increased above the minimum requirement, further changes in the properties occur, rendering them more similar to those of standard commercial leather. The tanning process is thus postulated as being to some degree two-stage in nature; although it does not exhibit a clear dividing line, the second stage by no means constitutes mere impregnation.

A number of papers followed dealing with the tanning action of new tanning materials or combinations of tanning materials. By way of an example, mention is made of the author's studies of the relationship between the constitution and tanning action of aromatic syntans or those dealing with the increase in hydrothermal stability due to combination of vegetable tanning agents with aluminium salts, aldehydic compounds such as tetrakis(hydroxymethyl) phosphonium (THP) salts, glyoxal, glutaraldehyde and oxazolidines[95], the latter also in combination with melamine resins or with protein condensates. By contrast, Heidemann dealt with aspects of tanning theory in a more generalizing summary by addressing the question of how the masses of tanning materials and fatliquors are deposited in the collagen structure. He also compared chrome and vegetable tanning according to the different bond positions on the collagen.

In accordance with the great economic significance of chrome tanning, which despite numerous studies was described by E. Brown of the Eastern Regional Research Center (ERRC) of the U. S. Department of Agriculture as being still more art than science[96], a whole series of theoretical tanning studies have been developed to the subject. Based upon sequence and structure data of the collagen, Heidemann calculated the scale and location of the (clustered) binding of the basic chromium sulphate complexes. Reich et al. and Covington et al. addressed the question of the number of nuclei in the bound chrome complexes, employing such modern and powerful methods for the purpose as extended X-ray absorption fine-structure spectroscopy (EXAFS)[97]. A number of authors believed the time had arrived for computer modeling of (chrome) tanning. The limited computer processing power did, however, impose restrictions. Only a portion of the five triple helices of the Smith microfibril were analyzed, and with the exception of the final studies, no consideration was given to the water bound to the collagen. In summing up his own, somewhat modest results, Fennen stated that computer modelling reaches us that chrome tanning is possible. The rapid development of hardware and software in the interim has doubtless provided a better basis, and, following a long break, work on computer modelling of tanning processes has therefore resumed, including modelling of the telopeptide regions in consideration of the possible crosslinking (tanning) of collagen under the influence of transglutaminase[98,99].

Very recently, Covington et al. attempted to produce a unified basis for a theory of tanning[100]. The emphasis lay upon thermodynamic considerations, with T_s serving as the preferred or sole criterion of leather formation. Although Covington states that "the future of tanning lies with chrome salts", he also derives proposals for future alternative tanning processes from the concepts of tanning theory developed. Finally, Ramasami attempted to provide a new framework for Wilson's one-time dream of a universal theory of tanning by drawing attention, much as Covington had done before, to the need to consider the integral collagen-water-tanning material system in all theoretical considerations, and to differences in the influence of the individual tanning processes upon the collagen-water matrix. Bienkewicz and Heidemann and most recently Kellert and Manzo were among those who had already pointed out this fact, which had also been the subject of occasional experimental study; it had however generally been ignored in tanning theories before this time[101]. These most recent developments will be considered in descriptions of the current state of tanning theory.

The developments in tanning theory over the last two centuries, the present description of which lays no claim to completeness, show that their meaningfulness has increased in proportion to the developments in knowledge of skin collagen and of the tanning materials concerned and to the availability of new, more powerful methods of analysis. The new tanning materials have been considered with regard to tanning theory to a greater or lesser degree in accordance with their economic significance, the greatest attention consequently being paid to chrome tanning (Table 3.4)[102-104]. Tanning theory has now reached a stage in which the key points are validated, and therefore forms the basis both of practical action, and of research and training. It will be summarized below with specific reference to practical aspects.

Fundamental Collagen Chemistry in Leather Making

Table 3.4 Development of the concepts of tanning theory

Time frame	Issues in tanning theory (key words)	Scientists involved (selection)
2nd century A. D.	Astringency = tanning capacity	Galenus (The Physician)
18th century	Tanning is the formation of salt, and increases the weight of the collagen	Seguin
19th century	Tanning is the coating of the fibers, leatherlike drying-out is decisive effect	Knapp
20th century	Tanning follows the laws of adsorption	Stiasny
Mid-1930s	Tanning is dependent upon chemical interaction between tanning agent and collagen: a universal tanning theory is not possible	Bergmann, Küntzel, Stiasny, Wilson, et al.
1935 ~ 1950	Explanation of the chemical interaction in the chief tanning processes	Küntzel et al.
1950 ~ 1970	Detailed presentation of the concepts of tanning theory, based upon growing knowledge of collagen	Grassmann, Heidemann, Reich, Stather, et al.
Early 1990s	First efforts in computer modelling of the tanning processes	Working group at the ERRC (USA)
Since 1998	Consideration of collagen-water-tanning material as a system (the leather matrix); the role of suprastructures in the hydrothermal stability of the leather matrix	Covington, Ramasami, et al.
Present time	Consolidation of theoretical concepts, in particular in the area of chrome tanning	Working groups at the ERRC (Brown), BLC (Covington), CLRPI (Ramasami)
Future	Precise formulation: "What substance tans where and causes what. Redefinition of the function of tanning"	

3.3 Current Knowledge of the Nature of Tanning

A modern theory of tanning can be universal only in the sense that it initially focuses upon the effects common to all tanning processes. It must then be subdivided into the chemically diverse groups of tanning materials and their correspondingly diverse interactions with collagen. The theory should above all be practical, i. e. it should be geared to conditions in the field of research, and provide this field with a scientific basis for further developments in tanning agents and tanning processes.

The question thus first arises regarding the current role of tanning in the manufacture of leather. Its first function is now the manufacture of a semifinished product which is rendered suitable for partial dehydration and subsequent modification of the thickness by shaving and/or skiving, can be stored and transported in the damp condition, and equipped at a later stage with desired properties in wet finishing. Retannage, which to some degree is secondary, then completes primary tanning effect and guarantees, in conjunction with fatliquoring and mechanical softening by staking and milling, the manufacture of serviceable leather. The forms of leather most important today, in particular chrome-tanned and "free of chrome" (FOC) types generally tanned with glutaraldehyde (chrome-free leather) are therefore based on "two-stage" tanning of this kind[105]; shoe upper leather, furniture leather and automotive leather are relevant examples. The division of tanning into two stages simplifies salvage of the by-products of thickness processing of the semifinished products, and enables primary tanning to be performed by an unconventional process. Sulphur or waterglass tannage serve as examples here. Primary (or main) tanning thus becomes a de facto pretannage, and retannage becomes in purely quantitative terms the main tanning. Both together are subject to principles of combination tannage (Figure 3.1).

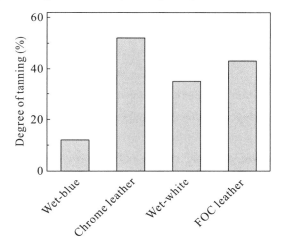

Figure 3.1 The level of tanning attained during (pre) tannage to wet-blue or wet-white and by its retannage in wet finishing to chrome or FOC leather

Thanks to our knowledge of collagen, it is now possible to define the nature of the two tanning stages in the light of the changes brought about on the collagen. Despite all the differences in desired leather characteristics, material structure, and reactivity of the tanning materials used, the two tannages in combination bring about stabilization of the collagen structure: tannage prevents the collagen microstructure from collapsing when the water is removed. This characteristic is accompanied by reduced swelling capacity, enhanced resistance to hydrolytic influences and to microbial and ultimately, therefore, enzymatic attack, and is often, but not necessarily characterized by high hydrothermal stability.

The effects described are permanent in the sense that they do not change significantly even under repeated moistening and drying of the intermediate product thus obtained. A further characteristic is that the tanning materials remain present in the leather. This constitutes a distinction from stabilizing of the fiber network by "self-crosslinking" of the collagen, triggered

Fundamental Collagen Chemistry in Leather Making

for instance by concentric shrinkage of the fibrils owing to dehydration with solvent or effected enzymatically, as in the case of transglutaminase treatment. In the first of these cases, the structural stabilization is not water-resistant without further measures (impregnation and/or hydrophobing), i. e. all characteristics described above cease to apply in the wet condition ("pseudo-leather"). This limitation probably does not apply to enzyme-driven self-crosslinking, assuming that it is achieved on any practical scale. This does not however make the enzyme a tanning material; it can however be regarded as a temporarily active substance which triggers tanning. This thus constitutes the basis for a modified definition of tanning:

Tanning is an industrial process stage during leather manufacture in which the collagen fiber network of the hides and skins is permanently stabilized in its chemistry and structure such that it no longer collapses and adheres to itself when water is withdrawn from it, i. e. it dries out in a leatherlike manner. The intermediate product thus obtained exhibits reduced swelling capacity in the wet condition, is less susceptible to hydrolytic and enzymatic attack, and frequently also exhibits greater hydrothermal stability. All substances which bring about these effects and become part of the leather are tanning materials, irrespective of their chemical structure and the nature of their bond to the collagen.

Whether all or only certain of the criteria stated are met in specific cases and whether they are met in part or in full is dependent upon the type of tanning material used, and also upon the other process stages, and is geared towards the desired leather characteristics. As formulated by Gustavson: "*Hence, the criteria given for the properties of tanned collagen and the definition of leather must be applied in due consideration of its intended use. Even in a strictly theoretical evaluation of the potency of a tanning agent, it is advisable not to emphasize any particular property or constant to the exclusion or detriment of any other criterion or criteria.*"

How the essential criteria by which the changes brought about by tanning on the collagen are to be assessed and characterized analytically in the light of current knowledge is described below.

3.3.1 Characteristics of Tanned Product: Cause and Demonstration

The changes brought about on the collagen by all tanning methods (though to different degrees) are described below in brief terms of their origin and the methods now available for their demonstration.

3.3.1.1 Leatherlike Drying-out

Prevention of the collagen structure from collapsing when the water is removed is identical to that which has been described since time immemorial as "leatherlike drying-out" and defined by Gustavson. The cause is the retention of air-filled voids between the collagen fibers and fibrils. As already explained in the introduction with reference to the work of Küntzel, this characteristic is due to the fact that the collagen fibrils do not adhere to each other during drying. They would be capable of doing so under the influence of the surface tension of the water owing to their flexibility and the presence of a number of adhesive groups (the meaning of the term: "collagen" is, after all, "glue-forming"). They are however prevented from doing so when they are stiffened and/or their reactive groups (those with the potential to adhere crosslink) are blocked by the tanning materials. It can clearly be seen that this may occur both by crosslinking, and by "distance enhancement" and mere coating.

Leatherlike drying-out, which anyone can learn to assess subjectively as the soft and warm handle, can also be defined numerically by a number of objective measures. All methods concerned are based upon the fact that isolation of the fibers and fibrils and the resulting porosity are measured directly or indirectly by way of the changes in characteristics caused by the porosity and/or are visualized (Table 3.5).

Table 3.5 Measurement of stabilization of the collagen structure by tanning

Characteristics indication of tanning/leather formation	Variables
Leatherlike drying-out	Handle, softness, thermal conductivity Porosity, apparent density, internal surface area Leather yield, yield on leather volume Visualization: optical microscopy, TEM, SEM, AFM
Thermal/Hydrothermal stability	DSC measurements Shrinkage temperature (T_s) Force measurement under isometric shrinkage Stress measurement
Swelling capacity	Water take-up/mass increase, including under the influence of variation in pH (charge swelling) or the addition of hydrotropes (hydrotropic swelling). Gravimetric measurement, measurement of the thickness and compressibility; dewaterability under pressure
Hydrolytic stability	Solubility in hot water, solubility in acids and alkalis, solubility in the presence of hydrotropes
Enzyme resistance	Degradation behavior under the influence of trypsin, pepsin, papain, effect of collagenase, composting test, mould infestation test

Porous drying-out gives rise to different volumes of tanned product than does parchmentlike drying-out. Indication of the volume rendement introduced by Russian authors as a measure of fiber isolation, expressed in cm^3/g of collagen, is therefore an illustrative quantity. It differs from one tanning method to another, and is dependent primarily upon the dosage of tanning material. It is, incidentally, very high in pseudoleathers (acetone-dehydrated pelt) (Figure 3.2).

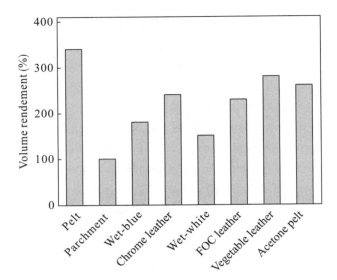

Figure 3.2 Volume rendement of pelt, dry semi-finished products (wet-blue, wet-white), various leathers, parchment and acetone-dehydrated pelt

Conventional microscopic images of dry leather produced by optical microscopy, SEM or AFM reveal most impressively the isolation of fibrils and fibers. Again, another example: if dried pelt and dried wet-blue are hydrated in the ESEM at 98% relative humidity, the images of originally parchmentlike dried-out pelt illustrate very clearly the homogeneous and indeed gel-like mass of the largely rehydrated collagen. Conversely, the isolation of chrome-tanned fibrils is completely retained under comparable conditions, and is even more convincing than dry images.

3.3.1.2 Reduced Swelling Capacity

Fully hydrated pelt exhibits a water content of over 200% in terms of collagen. The "free" water not directly bound to the collagen (*bulk water*) accounts for the greater part and fills the voids between the collagen fibers and fibrils. The collagen is held together by covalent crosslinking, electrostatic forces and hydrogen bonds determining the water content and thus the level of swelling. The natural crosslinks increase with the age of the skins: adult collagen thus swells notably less than young collagen. The same applies to the influence of crosslinking tanning: here too, the level of swelling and the water content are reduced. This does not become particularly apparent, however, until either the balance between acid and basic groups, which exists at the IEP and ensures minimum swelling, is disturbed, or the high number of intracollagenous hydrogen bonds with their stabilizing action is reduced. In both cases, the degree of swelling rises considerably since, due in a sense to structural destabilization of the collagen, more space has been created for further take-up of water. The former case applies to acid or alkali swelling; the latter is responsible for hydrotropic, lyotropic swelling. Tanning is capable of reducing both swelling effects. Gravimetry is the most convenient method for measurement of the different water take-up levels.

3.3.1.3 Increased Hydrolytic Resistance

As has already been stated, "collagen" means "glue-forming", i. e. untanned collagen can be broken down by treatment with hot water and placed in solution. This characteristic is eliminated to a greater or lesser degree by tanning, resulting in the case of boil-resistant chrome leather to complete resistance to hot water. The crosslinking action of the tanning materials is responsible for this characteristic. The boil resistance is determined gravimetrically as described by Gerngross. The measurement result reflects by definition the thermal stability of the tanning material-collagen bond. It is lower in the case of hydrogen bond formation than in the case of covalent bonding. By a shift in pH away from the IEP, the conditions of hydrolysis can be intensified. For example, an alkali resistance can be measured which is likewise in proportion to the tanning material.

No limits are placed upon the experimenter's imagination for the design of such degradation tests: even the hydrotropic influences upon the placing in solution may be measured. As will however be appreciated, the results which may be gained from such methods are largely identical to those obtained from simpler swell measurements.

3.3.1.4 Increased Resistance to Microbial and Enzymatic Degradation

Ever since the effects of tannage were defined, the resistance to putrefaction has been one of the most important. It is seldom considered however that this is a feature of wet leather. Leather in this state is however a temporary exception. The improved resistance to microbial, i. e. enzymatic attack, is on the one hand a consequence of the fact that leather, in comparison to hide, contains incomparably fewer soluble substances which serve as a potential culture medium for microorganisms. On the other hand, crosslinking, as evidenced by the reduced swelling (see above), brings about a delay in the attack, with the effect ultimately of sterically obstructing the enzyme attack. In addition, the tanning materials also act in part as enzyme inhibitors and/or biocides. Ramasami et al. were able to demonstrate recently that the splitting of model peptides frequently employed for measurement of collagenase activity (2-furanacryloyl-leucyl-glycyl-prolyl-alanine) are inhibited by Cr(Ⅲ) complexes, and incidentally by binuclear and trinuclear complexes more strongly than by tetranuclear ones. Direct binding of the chrome complexes to the enzyme was demonstrated by gel electrophoresis. As the leather contains up to 1500 ppm of unbound Cr(Ⅲ) complexes, the reduced degradability of chrome leather requires no further explanation. It may be assumed that other enzymes are subject to similar influences and that their effect is reduced or eliminated altogether. A finding of Grassman's working group is of interest to tanning theory with regard to the degradation by collagenase: the polar amino acids, in particular Glu and Asp, are enriched in the non-degradable residues of chrome-tanned collagen in addition to the chrome tanning material. This constitutes independent proof of the already accepted binding of the chromium complexes to these two acid amino acids.

Due to the assumed significance of leather's resistance to putrefaction, Thomas and Seymour-Jones introduced an enzymatic laboratory test as early as 1924. As collagen is resistant to proteases with the exception of collagenases, the rates of degradation, in some cases very high, of untanned collagen observed with trypsin, pepsin or papain may be understood only if the test conditions are taken into consideration. These do in fact often contain components, pH or temperature conditions, which bring about parallel hydrolytic degradation of the collagen. It is then only logical to conclude that this collagen, already damaged, is also degraded by proteolytic enzymes, i. e. the degradation may at best be delayed by a tannage. In other words, resistance to enzyme attack reflects in principle nothing other than swelling capacity and hydrolytic resistance. In any case, any skilled person in the field is aware of the susceptibility of damp leather (e. g. wet-blue) to microbial attack and of the decay of leather in soil.

The increased resistance to microbial attack is a notable characteristic of leather and can be influenced by tanning: only, however in the progress of degradation, and not in its ultimate scale. Test methods must be considered critically; tests under practical conditions, geared for example to mould attack on wet-blue and its prevention, are of greater relevance. It is regrettable that no reliable data exist for the susceptibility of leather to degradation in standardized composting tests.

3.3.1.5 Increased Hydrothermal Resistance

The increase in hydrothermal stability, i. e. the rise in the denaturing temperature of native collagen, has always been regarded as the characteristic indicating completion of tannage by crosslinking. It is undisputed that the T_s which is based upon it is a quantity of interest in tanning theory[1]. It was first measured by Powarnin and interpreted in the context of tanning theory as early as 1910. It provides an indication of various types and grades of crosslinking, and its isometric measurement yields considerable information on the modification which the collagen/leather structure has undergone. It provides little indication of the practical value of the leather, as (wet) leather is seldom subjected to thermal load during processing and in use. The thermal load during shaving is also lower than is often assumed. Attainment of as high a T_s value as possible is also not therefore an objective in the development of new tanning methods, a fact already pointed out by Gustavson: *"Thus, characterization of tannages mainly from the point of view of the degree of hydrothermal stability attained in the final product by a certain tanning agent may lead to erroneous conception, or even invalidate the conclusions drawn from accentuating one property at the sacrifice of the other criteria."*

Despite this imperative qualification, certain considerations from the perspective of tanning theory are appropriate, particularly since Covington et al. attempted to formulate a new tanning theory encompassing all tanning processes on the basis of detailed thermodynamic discussions.

Fully hydrated (> 200% water/collagen), native collagen undergoes denaturing when heated to approximately 62 ℃, which can be observed by shrinkage of the samples to a third of the original length, and by the peak in measurements taken by means of differential scanning calorimetry/differential thermoanalysis (DSC/DTA)[106]. The area below this peak corresponds to the heat requirement of the endothermal melting process, the location of the peak maximum to the T_s (Figure 3.3). DSC measurements are now the preferred method, as they permit measurements on samples with any given, constant water content in a closed measurement vessel, require small samples, and are fast and accurate. They also deliver the desired caloric data for thermodynamic analysis.

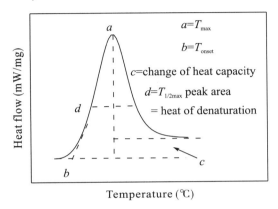

Figure 3.3 Schematic diagram of the characteristic produced by DSC measurement and the relevant quantities of measurement and operands

The melting process is the transition of the triple helixes of the native collagen to the statistical coils of the shrunk collagen. This process begins in the C-terminal end regions of the a chains, which are free of stabilizing hydroxyproline, and continues initially as a rating-out of the chains from their triplex, which may be described as "uncoupling by zip off"[107]. The transition to the coil structure then follows. These processes do not take place continuously over the entire collagen molecule, but in blocks, or "co-operative units", of differing stability, section by section at different temperatures, as demonstrated by Primakov on collagen solutions. The stability is dependent upon the content of polar amino acids ("non-ordered" regions of the triple helix), and also upon the content of structure-stabilizing hydroxyproline in the individual regions of the molecules ("domains"). Native, fully hydrated collagen (220% H_2O) contains 52 "co-operative units", which incidentally corresponds closely to the number of bands of highly resolved cross-striation in the TEM, which is also a reflection of the polar regions. The number drops to 4 for total dehydration (220% H_2O). The cause is the "self-crosslinking" of the polar groups of the collagen, in which the individual "co-operative units" combine to form larger blocks. The crosslinking action of tanning materials is similar. The number of "co-operative units" for chrome-tanned collagen for example is reduced to 19 in the wet and 3 in the dry state.

The T_s is at its highest (210 ℃) in water-free untanned collagen. This is related to the fact that all reactive groups of the collagen crosslink within the collagen when water is removed from the collagen-water system. As described above, crosslinks introduced by tanning increase the test of wet collagen as anticipated, but are more of a hindrance to self-crosslinking within the collagen when water is removed: in the dry state, tanned collagen has in fact a lower T_s than untanned collagen.

Let us further consider the T_s of fully hydrated collagen or leather. It may be influenced in a number of ways: it is reduced by lyotropes and other compounds which promote swelling. This is logical: collagenous bonds are broken; organic hydrotropic substances (e. g. urea) have this effect, as do a number of simple salts, acids and bases. Conversely, certain simple salts, e. g. sodium sulphate (see below), increase the T_s slightly under contain circumstances. As the effect no neutral salts cannot be interpreted as a hydrotropic effect upon collagen, however, since they do not break its bonds. Their action can be explained only by the assumption that they influence the water envelope of the collagen. A contain analogy exists to the effects known in protein chemistry of salting-in and salting-out with neutral salts. For example, NaSCN belongs to the former of these groups, and NaF, Na_2SO_4 and $Na[N(CH_3)_4]$ to the latter.

Tanning agents generally raise the T_s specifically to the region of 70 ℃ ~ 85 ℃. Only by chrome tanning with basic chrome sulphates or certain combinations of tanning agents is it raised to 110 ℃ ~ 115 ℃. Chamois tanning incidentally reduces it to 50 ℃. The basis for Covington's deliberations on a universal tanning theory was now the origins of the differences in T_s, and whether anything could be learnt from them regarding the nature of tanning. Based upon thermodynamic considerations, he concluded that the measured differences in test could be attributed to the different ratios of entropy to enthalpy component in the individual tanning types, i. e. the structure-forming effect of the individual tanning materials upon the "co-operative units". These variables are accessible for analysis through isothermal measurement of the shrinkage progression. As already performed in the classic study by Weir, or by direct measurement of the enthalpy of fusion by DSC[108, 109]. This still fails to explain the chemical and structural origin of the differences between the different tanning methods. Covington proposes the following explanation based upon the special status of chrome-tanned leather.

In contrast to chromium chloride or chromium persulphate, which yield a maximum test of 85 ℃, chromium(Ⅲ) sulphate yields T_s of 115 ℃. If the sulphate is removed by intensive elutriation, the test drops to approximately 85 ℃. If sulphate is added again, the T_s rises once more to 115 ℃. The clear influence of the sulphate anion cannot be attributed to the three initially conceivable arrangements of the sulphate: a) component of the chromium complex (excluded by EXAFS measurements), b) counterions to the chromium complex, or c) counterion to dissociated amino groups of the collagen. In the latter two cases, the bond forces are too low to explain the observed differences in test. An influence is thus postulated of the complete matrix comprising collagen, bound chrome tanning material and water, which will be termed here the "leather matrix", such that the sulphate groups stabilize the suprastructure of the water envelope. Sulphate does in fact belong to the salts mentioned above which increase the T_s. Based upon the results of other authors, Covington's points out that a subsequent treatment of

Chapter 3 The Theory of Tanning—Past, Present, Future

collagen already treated with chromium perchlorate with sulphate or even with certain organic anions likewise brings about an increase in T_s to >115 ℃.

How does the sulphate now act within the suprastructure of the leather matrix? Covington presented the following for discussion: a) a stabilizing influence of the water structure itself, b) intensification of the binding of the water to the collagen, and c) increased binding of the water to the chromium complexes. All three conceivable modes of action may also act in combination. In the same way, Covington interprets the synergetic effect of a combined application of vegetable tanning materials with metal salts or aldehydes, which also leads to values of T_s of >115 ℃, as a stabilization of the leather matrix. It therefore represents in a sense the organic counterpart to the action of sulphate in chrome tanning. The high T_s (>150 ℃) of calcified collagen (bone) noted by Kronickea suggests much the same[110]. All types of tanning lacking such an additional stabilization of the matrix reach T_s of only 70 ℃ ~ 85 ℃; where lyotropic effects are also present, the T_s drops below that of the native collagen ("negative tanning").

This is indisputably an original approach in the context of tanning theory, the most important aspect of which is that of inducing the water always associated with the collagen in the discussions. Covington himself points out, realistically, that a great many questions remain to be as answered and must be addressed experimentally. With regard to the hydrothermal stability of leather, it must also be mentioned that the T_s of the leathers is reduced in the course of its natural ageing, in particular under exposure to thermal stress, and that it drops to the level of untanned collagen.

A hydrothermal impact which ultimately leads to shrinkage naturally also brings about a constant change in volume of the leather matrix as the leather is heated. Studies of this phenomenon by Ramasami et al. reveal the complexity of the processes, an initial increase in volume in the range 26 ℃ ~ 51 ℃ being followed by a reduction, which then becomes substantial at the T_s (note: the change in volume during melting/at the T_s is in no way comparable in scale to the drastic change in length)[111]. The methods employed in the studies are interesting and warrant continuation as intended.

Changes in the surface area are also a consequence of partial or complete denaturing of the collagen. The absence of these phenomena in the boiling test on chrome leather may be a suitable means of implant verification of proper tanning, but no more than that. This is not the case with the shrinkage of essentially dry leather under varying temperature and atmospheric humidity conditions (cyclical climatic testing), which is important for instance for the assessment of dashboard leathers. Substantial differences exist here between chrome and chrome-free tanned leathers. These differences await theoretical elucidation. If the values for the above parameters found throughout the literature are summarized for the main tanning methods, it can be seen that although different effects are observed; only limited information is gained (Table 3.6).

Fundamental Collagen Chemistry in Leather Making

Table 3.6 Comparison between the tanning criteria for wet leather (approximate values based upon a number of different publication)

Variable	Pelt	Chrome leather	Vegetable leather	Aldehyde leather
T_s	62 ℃	>110 ℃ (chromium sulphate tanned)	75 ℃ ~ 85 ℃; >110 ℃, if crosslinked by synergistic action	75 ℃ ~ 85 ℃
Swelling	100%	50%	35%	60%
Solubility in hot water (0.1 mol/L NaOH, 65 ℃, 1 h)	(85% dissolved) 100%	36%	40%	9% (CH_2O)
Resistance to enzymatic attack (trypsin)	(73% dissolved) 100%	6%	40%	2% (CH_2O)

3.3.2 Reactions of Tanning Materials with the Collagen

A condition for the features of tanned product described above is an interaction between the tanning materials and the collagen. A number of possibilities exist for investigation of the nature of this interaction. For ease of comprehension, the possible interactions on the collagen will be considered briefly.

3.3.2.1 Possible Interactions on the Collagen

The reactivity of the collagen is determined by the functional groups present within it and their accessibility in the various regions of the collagen structure. The bound quantity of collagen is determined primarily by stoichiometric aspects, but is also influenced by available surfaces on the fibrils, elementary fibers and fibers, and in the voids between them (capillaries, pores). The tanning materials exploiting these potential reactions on the collagen and the mechanisms by which they do so is now well understood at least in basic terms. Brief reference to the following summary is sufficient in this case.

The fact that we now know so much about this, the chemical side of the tanning material-collagen interaction, is partly due to the fact that a plethora of independent methods exist for demonstrating whether or not interactions have occurred. Their application also permits classification of new or future tanning materials, and is a welcome basis for screening (Table 3.7). "Interaction" primarily means nothing more than the creation of a compound involving the tanning material in question and the collagen. The fact that crosslinks are created in the process can be inferred from the stabilizing effect upon the collagen as a pre-condition for leatherlike drying-out and from the increase in T_s. The constitutional formulae frequently encountered for tanning reactions are illustrative and possibly even largely correct, but do not constitute evidence in the strict chemical sense.

Chapter 3 The Theory of Tanning—Past, Present, Future

Table 3.7 How to characterize the tanning material-collagen interaction

Object of characterization	Means of characterization
Tanning process proper	pH dependency, quantity of (available) tanning material, time requirement
Leather matrix	General: IEP shift, Δ acid-base bond, ΔT_s
Model substrates based upon collagen	Spectroscopy: FT-IR, Raman, NMR (^{13}C, ^{27}Al, H), EXAFS
	Microscopy: SEM, TEM, AFM (cross-striation); EDX
"Collagen-like" model substrates	Molecular disperse collagen solutions (assembly), gelatine solutions and gels, hide powder, collagen film, catgut
Modified collagen	Model peptides, polyvinylpyrrolidone, polyamides, ion exchangers
	Blocking and introduction of functional groups

Precise demonstration would necessitate isolation and identification of defined collagen fragments of the form amino acid-tanning material-amino acid. Unfortunately, this instrument cannot be employed owing to the fact that the required hydrolysis of the collagen breaks the collagen-tanning material bond at the same time. Little direct evidence of crosslinking exists. The "chemical circle" is the classic example, showing as it does that crosslinking is successful only when the relevant spatial requirements are met, i. e. intervals of the reactive groups of the collagen and the tanning agent are "compatible". Similarly, conversion with sulphochlorides and of quinone was studied. In the case of chrome tanning, model studies must suffice, such as those produced by Heidemann is based upon the sequence and structural data of the collagen, or those forming the basis of the computer model to which reference has already been made. By no means has exact proof been obtained to date of the regions within the collagen structure in which binding of the tanning material actually occurs, and of whether its action is always that of crosslinking.

Critical analysis of the current state of tanning theory as described here thus immediately reveals a series of gaps. Even certain facts assumed to be validated may now be challenged owing to the improved methods of investigation. This also applies to chrome tanning, which has been studied so comprehensively, a fact pointed out recently by Covington. Development of a more perfected theory of tanning remains a task for the future. Selected considerations regarding such a theory follow.

3.4 Tanning Theory in the Future

A universal tanning theory will continue to require agreement on the fundamental changes to the properties of collagen brought about by tanning. However, many ways will exist by which this aim may be achieved. Each new tanning process while therefore requires its own specific principles of tanning theory. The established tanning processes, however, have by no means been fully explained in terms of their interacting fairs, not even in terms of their purely chemical processes. This can be demonstrated by way of three aspects of three major tanning processes:

(1) What is the basis for the considerable differences in the tanning action of the various vegetable tanning materials, such as certain benefits of the tanning material tara, and what is the function of quinoid structures in the binding of vegetable tanning materials?

(2) Do aldehydes react as monomers, or as polymers?

(3) Chromium complexes of what size are responsible for tanning, and where do they take effect? Are chrome tanning materials bound other than to the carboxyl groups of aspartmatic acid and glutamic acid, and what is the actual function of the residual sulphate?

Whether it is necessary or desirable in the practical interests of leather manufacture to find answers to these particular questions and many others in tanning theory and whether time, financial resources and expertise can be found for their investigation is another issue. We can be certain however that the increasingly refined study techniques will one day enable answers to be found.

Generally, it may also be stated that in the future, tanning theory will have to give greater consideration to steric aspects of collagen-tanning material interaction than has been the case or even possible in the past. In other words, in what regions of the collagen structure do the tanning materials bring about the desired stabilization of the collagen? Do they have these effects in the intrafibrillar/intermolecular region, or do they take place in addition and possibly to an even greater degree in the interfibrillar region of the elementary fibers and the fibers? As the intrafibrillar region of the collage is not readily accessible directly for observation, assembly experiments with collagen solutions for characterization of collagen-tanning agent interactions, such as those described by Tuckerman et al. and Ramasami et al., are increasingly significant.

Another unanswered question is that of the ideal particle size to tanning agents, which also includes their capacity for diffusion. Particle size should not be understood here as a static dimension: tendencies towards aggregation must also be considered, as should other conceivable changes in form which influence the capacity for diffusion, such as the way in which large biomolecules extend in order to pass through membrane channels of small diameter. This is particularly relevant for polymer tanning agents, which are becoming increasingly important. Owing to the rapid developments in hardware and software, computer modelling will be able to contribute substantially to clarification of this issue.

All structural elements of the collagen possess a water component which is bound to a greater or lesser degree. Any form of attachment of tanning material requires an interchange of sites, with tanning material taking the place of water. Alternatively, it may be more accurate to describe collagen, tanning material and water as forming an integral suprastructure, as described above. Should this be the case, what are the practical consequences, and how can this question be answered by experimentation? This area represents a further key subject of future research.

Leaving aside synergistic effects, all tanning theories developed to date pay scant attention to combination tanning processes. With the exception of a few brief chapters in the relevant

Chapter 3 The Theory of Tanning—Past, Present, Future

literature, the last major survey dates back to 1936. Yet it is combinations of substances which promise interesting new effects owing to their chemical and structural variety. Both experimental and theoretical research is required for this purpose, however. The natural, enzyme-driven and age-dependent crosslinking of collagen also constitutes a form of tanning. Development of this mechanism into an industrial tanning process by the *in-situ* of enzymes is an old proposal, one recently made again several times with the appearance of suitable enzymes at affordable cost[112].

New analysis procedures must of course be tested constantly for their suitability for application in the area of tanning theory. ESEM and AFM constitute examples from the field of microscopy[113]. The application of EXAFS, NMR and small-angle X-ray scattering (SAXS) offers further potential[114-116]. Further development of tanning theory is also desirable for more practical reasons: as far as the main leather types are concerned, tanning has mutated to a form of retannage, which permits only processing of the thickness and is intended to provide the semi-finished product with a degree of storability. This immediately gives rise to two questions, however. Firstly, need such pretannage really be performed by means of traditional tanning agents (chrome, wet-white)? Attempts at sulphur tannage or water-glass treatment are emerging as alternatives and are of interest from a recycling point of view. Do further alternatives exist for structural stabilization of the collagen, initially only temporary? The semifinished products require further filling and softening? This is achieved by retannage and fatliquoring, which are performed during wet finishing and determine the product's characteristics; they have therefore grown in importance. Do these, however, need to be two substance groups? What additional functions do "lubricating tanning agents" have which are not shared by compact products such as Microsan? This also represent an area of theoretical and experimental research.

As these selected examples show, further development of tanning theory is both necessary and possible. Greater understanding of the nature of leather formation, of which tannage will always be the chief substep, will not only ensure that the performance properties of leather remain in step with ever stricter requirements, but will also enable careful tanning material design to rationalize processes and overcome ecological problems.

Chapter 4 The Significance of Water for the Structure and Properties of Collagen and Leather

4.1 Introduction

The present study follows previous studies of the "Structure and reactivity of collagen" and "Physical processes on collagen" and "The theory of tanning". Its purpose is to provide a systematic survey with regard to the significance of water for the structure and properties of collagen and leather, given that current knowledge is comprehensive but by no means complete, is not free of contradictions.

The relationship between the structure of collagen (and leather) and its properties is only one particular instance of the protein-water interaction in general, and at the same time is a consequence of the unique properties of water. This relationship, mysterious and still not yet fully understood, is adaptable in its manifestation, versatile, and extremely influential. Water is the center of all bioprocesses, and in this sense is thus also involved in the origin of collagen, which is incidentally a very old substance. It is now regarded as a fact that proteins owe their essential structural properties, functions and effects to their water content. Owing to this particular significance, the interaction of protein and water has been studied comprehensively, numerous study methods being available.

All processes taking place on the collagen during its conversion to leather also influence the water balance of the collagen, with far-reaching consequences. Water is at one and the same time a crosslinker, spaceholder, filler and softener. It combines within itself functions which are subsequently assumed by tanning materials, fillers and fatliquors, these substances in some cases displacing the water. If purposeful product developments are to be placed on a sounder theoretical footing, with the objective of enhancing leather properties and rationalizing process steps, a sound theoretical basis must be found for the existing empirical knowledge of the influence exerted upon the collagen and leather properties by the interaction of collagen, water and leather-forming ingredient (in particular tanning materials themselves, fatliquors and fillers). Such a theoretical basis also facilitates computer modelling of the transformation of collagen into leather.

The significance of water for the structure and properties of collagen and leather, and for all processes of conversion from (hide) collagen to leather[117], can hardly be overestimated. As far as

Chapter 4 The Significance of Water for the Structure and Properties of Collagen and Leather

leather manufacture is concerned, it needs only be considered that with hide fresh from slaughter, the process begins with a material the principal component of which is 65% ~ 70% water, and completes it with a material of which water forms only 10% ~ 15%. Leather manufacture may thus also be defined as the removal of water from a biological material without destruction of its fundamental structure and without causing the microstructure to collapse.

Furthermore, water is the solvent for all substances involved in the transformation of collagen to the leather matrix. Numerous technical terms draw attention, for good reason, to the significance of the water: wet processes, wet finishing, wet end, wet blue, wet white, etc.

The interaction between water and leather and the water content of the leather are factors of importance for the leather properties. The unique comfort properties of the leather, is high water (vapor) sorption capacity with retention of its dry handle, and its active breathing characteristics reflect this fact, as does the dependence of numerous physical properties upon the water content. When undesirably high water sorption and permeability of leather must be countered by hydrophobing, water is the factor necessitating action. The dependency of leather's mass and dimensions, and in fact all physical properties up to and including its handle, upon the water content, and the serious mistakes which may be made by unskilled removal of the water during drying, have long been known.

The great significance of water for leather is shared by collagen. The latter's structure, function and properties, like those of any protein, may be understood only in the context of an integral protein-water system. The molecular structure of the collagen is stabilized by water, the fibrillar structure dilated or constricted as a function of the water content, with consequences for the diffusion and binding processes occurring during leather manufacture. Without water there are no ion charges, with their effect of attraction and binding. The water content has a bearing upon the thermal and hydrothermal stability of collagen. Swelling and depletion processes are used to control the leather properties through the intensity of opening up of the hide, and without the "*bulk water*" in the macroscopic collagen structure, there would be no diffusion processes, and thus no material transformation on the collagen.

All the above aspects and others raised in the course of the project have been taken into account in the significance of water for the structure and properties of collagen and leather, as well as for the processes of leather manufacture.

4.2 Collagen and Water: An Integral System

"*All things out of water are created.*"

—J. W. Goethe. Faust Ⅱ

Goethe did not have collagen in mind when he wrote these words. But are they not also true of collagen? The α chains with the propeptides, which promote solubility in water, are "discharged" in aqueous solution. Enzymes in solution in the water cause these propeptides to be cleaved off, hydroxylate part of the proline and lysine, and bring about crosslinking. The water is influential in the folding of the collagen molecules to form the triple helix and in their charge-driven assembly to fibrils. Later, it forms a mantle over collagen molecules, fibrils and fibers. H_2O permits dissociation of the collagen, acts as a solvent and transport medium for a number of substances, and is therefore a precondition for the reactivity of the collagen. Water is thus an integral part of collagen, as with all proteins. This particular pair of substances is, however, capable of particularly dynamic and varied transformations: as a fibrous protein equipped with voids of different dimensions (pores, capillaries) between the individual structural elements (molecules, fibrils, fibers, fiber network), it contains not only bound water of varying bond strength, but also *bulk water* in great quantity, up to 2.5 to 3 times its own weight. This collagen-water system is subject to continual influence during the various processes for transformation of collagen to leather, by changes in pH value, the action of salts, hydrotropes, tanning materials, etc. The result is a constant change in the ratio of water to collagen, e. g. towards a maximum and minimum water content during swelling processes and leather drying respectively. Even though the unbound water component generally changes first, the bound water components are often also affected. These phenomena all have far-reaching consequences for the structure. Not only that: when the leather is in use, the collagen is subject to the influence of water and water vapor in a number of ways, and reacts again with changes to the system and consequently to its properties.

4.3 The Consequences for the Collagen Structure of a Change in Water Content

Removal of the water from its compound with collagen has dramatic consequences for the collagen structure. When the capillary water and the loosely bound water are removed from untanned collagen by (gentle) air drying, the fibrils, which are still "coated" by firmly bound water, converge until they are completely and densely packed. The water content drops to approximately 15% ~ 20% of the dry weight of the collagen, and the material dries out in a hornlike manner (parchment). If the fibrils are prevented from converging in this manner, either by tanning of the collagen (stiffening of the fibrils), solvent dewatering or freeze-drying, the capillaries and pores become filled with air and the material is soft and porous. The scale of the fibrillar water envelope remains largely unaffected by this process, however. These latter two materials can be fully rehydrated, albeit at different rates.

The situation is quite different when the water content is reduced beneath this critical value of approximately 15% by substantially harsher drying conditions. The convergence of the fibrils triggered in this case leads to intracollagenous crosslinking, which is only partially reversible, if at all. If the molecularly bound water (H_2O Type I) is also removed, then strictly speaking no

Chapter 4 The Significance of Water for the Structure and Properties of Collagen and Leather

intact collagen remains, and the damage to the collagen is irreparable.

4.4 Water Determines the Collagen Characteristics

The native Type I collagen of the corium is fully hydrated. It owes its structural stability to the hydrogen bonds, the hydrophobic and van der Waals bonds, and the crosslinking, which is a function of age and is often somewhat neglected in discussions; also and in particular, to the mutually attractive positive and negative charges of the dissociated basic and acidic groups in the sidechain position of the diaminocarbonic (Lys, Hyl, Arg, His) and aminodicarboxylic (Glu, Asp) acids. The "native" hydration level in the sense of water content, including that of Type III (*bulk*) water, is also determined by this mechanism. If this structural stability is disturbed by external influences, such as a change in pH value, the addition of salts or a number of other substances, and/or changes in temperature, the level of hydration of the collagen also changes. Should this be expressed by swelling, it is always associated with an increase in the water content and in volume (Δ fiber thickness, length). The latter may lead to strain in the collagen fiber network, which in turn leads to plumpness, i. e. enhanced resistance to compression. The reverse effects are also possible: partial dewatering, with a collapse of the fiber structure. Swelling processes can be induced by changes in pH value (charge swelling), or by hydrotropic substances (hydrotropic swelling), which have different consequences for the collagen.

In the case of temperature influences, a completely different effect is at work, namely that of denaturing, i. e. the progressive transition of the collagen triple helix to the form of a statistical coil. This can best be observed on collagen solutions by modification of the denaturing temperature (T_d) and measurement of the optical rotation or of the circular dichroism. In the fiber composite, denaturing brings about a reduction in the shrinkage temperature (T_s), in extreme cases in conjunction with an entering into solution, i. e. with greater or lesser degradation of the collagen. Unlike the swelling processes, which essentially concern changes to the Type III water content, water Types I and II are always involved in the denaturing processes.

All phenomena have been the subject of theoretical and experimental attention in tanning chemistry, in the case of some phenomena for a long time. The reader is reminded of early attempts to explain the swelling processes by the Donnan theory of membrane equilibrium. This behavior of the collagen in response to physical and chemical mechanisms is certainly relevant, both to the manufacture of leather and to its characteristics, and is no less important than the chemistry and structure of the collagen.

4.5 Thermal and Hydrothermal Stability of Collagen and Leather

Native Type I collagen of the corium undergoes drastic structural change at 60 ℃ ~ 62 ℃, its T_s: a collagen or leather sample exposed to this temperature even for only a relatively brief period shrinks to a third of its original length, at the same time becoming thicker and acquiring an almost rubberlike elastic quality. Should the sample be fixed to prevent it from shrinking ("isometric" conditions), considerable tensile forces are generated in the sample which can be measured and analyzed in various ways. This shrinkage phenomenon is a manifestation of denaturing of the collagen, i. e. of the transition from the triple helix to the form of the statistical coil.

An understanding of the denaturing and shrinkage processes on the collagen, which incidentally are strongly influenced by the water content, is of great importance for many properties of collagen. Dissolved collagen molecules also undergo a helix-coil conversion, i. e. denaturing, when the aqueous solution is heated. This conversion can be measured by circular dichroism (CD peak at 225 nm), by a change in the specific optical rotation $[\alpha]_D$ and by the change in hydrodynamic data such as viscosity, sedimentation coefficient, and the results of light scattering measurements. The collagen's conversion from the native to the denatured state also modifies its accessibility for enzymes and antibodies.

Dissolved Type I collagen molecules of the corium of mammals denature in aqueous solution at a denaturing temperature T_d of 38 ℃. The denaturing temperature of other collagens, particularly fish collagens, is substantially lower, as a function of the hydroxyproline content. A relationship exists with the temperature of the aqueous environments in which the fish live.

The T_d of dissolved collagen can be measured easily and accurately. A water surplus is always present which undergoes no change during measurement. The situation is more complicated where a fibrous network of corium collagen is concerned. Owing to the networking of the fibrils and fibers and the crosslinking arising during growth/ageing of the collagen, the denaturing temperature is regularly some 24 ℃ above that of the dissolved collagen molecule. In this case, the T_d is termed the shrinkage temperature T_s, as denaturing is manifested in this case in shrinkage of the sample. The T_s is generally measured in the presence of a water surplus which does not change during the course of the measuring process. With a defined temperature increase over time (2 ℃/min), the T_s is measured reproducibly. The corresponding test specifications are standardized according to the official method recommended by the International Union of Leather Technologists and Chemists Societies (IULTCS), namely, IULTCS/IUP 16.

Shrinkage also occurs when the temperature is held constant for a longer period below the actual T_s measured as described above. Here too, a water surplus can be and is generally employed. Under such "*isothermal*" conditions, gradual shrinkage takes place which can be measured by means of a cathetometer in terms of the change in length of the sample. The kinetics of the denaturing process can thus be analyzed. The thermodynamic relationships applicable to the corresponding chemical reaction processes may therefore be applied. The half-change value of the shrinkage can thus also be calculated, the level of which corresponds, as expected, to that of the T_s.

The circumstances are different when the sample is prevented from shrinking by being

Chapter 4 The Significance of Water for the Structure and Properties of Collagen and Leather

placed firmly under tension during heating. Under such "isometric" measurement conditions, forces arise which can also be measured and analyzed mathematically ("hydrothermal isometric tension": HIT). From a purely metrological perspective, the isometric conditions can be created in various ways. A water surplus and fully hydrated samples, which though possible would be difficult to achieve in practice, are not generally employed in this case; instead, conditioned samples (free of Type III water) and conditions of defined varied atmospheric humidity are used (Table 4.1).

Table 4.1 Methods for measurement of collagen denaturing

Principle of measurement	Presence of water	Provides information on
Change in length of an unrestrained sample with increase in temperature	Enveloping water bath as heat medium	T_s
DSC	As desired, with closed vessel	T_s, enthalpy
Change in length of an unrestrained sample at constant temperature as a function of time (isothermal measurement)	Enveloping water bath as thermostat	Shrinkage kinetics
Force generated by a tensioned sample (isometric measurement, "HIT")	None, climatic chamber, water cell	Force generation
Stress measurement	Climatic chamber	Force generation
Stress/strain characteristic of shrunk, rubberlike samples	Fully hydrated samples	Interval between crosslinks

4.6 The (Hydro) Thermal Stability of Collagen and Leather as a Function of the Water Content and Tanning

Many studies have established that the T_s is dependent upon the water content of the collagen or leather samples[118]. All available results show that the lower the water content, the higher the T_s. The values measured for leathers subjected to different tannages differ accordingly (Table 4.2). The water content must be kept constant during measurement. DSC measurements should thus be performed in a sealed pan under a hot-stage microscope with sealed melting-point tube.

Fundamental Collagen Chemistry in Leather Making

Table 4.2 Relationship between the T_s, the water content and the tanning method.

Substrate	Percentage water content	T_s(℃)
Collagen	4	155
Collagen	19	121
Collagen	30	100
Collagen	98	61
Collagen	151	61
Chrome-tanned leather	0	160
Chrome-tanned leather	20	120
Chrome-tanned leather	150	115
Vegetable-tanned leather	8	140
Vegetable-tanned leather	19	122
Vegetable-tanned leather	35	87
Vegetable-tanned leather	124	85
CH_2O-tanned leather	0	160
CH_2O-tanned leather	20	118
CH_2O-tanned leather	150	75

When the T_s is measured under surplus water conditions, the temperature rise and associated input of energy into the collagen-water system progressively induce break-up of the stabilizing hydrogen bonds, the collagen molecule, the collagen-water matrix, and of course of the water itself; this is the root cause of the helix-coil transition[119]. An exaggerated description would be that under such measurement conditions, water is itself a hydrotrope, at one and the same time the trigger and the final stage of denaturing. The T_s does indeed rise when water is "diluted" by the addition of other water-miscible solvents or indeed when the T_s is measured in non-aqueous solvents. It must however be considered that counter-effects exist according to the concentration; the addition of up to 10% ~20% ethanol to the water for example results in the T_s falling, of higher quantities in it rising.

The type and concentration of the electrolytes (acids, bases, salts) and hydrotropes present also have a substantial influence upon the T_s. In general, a direct cause-and-effect relationship exists between swelling and T_s: an increase in swelling reduces the T_s, and *vice versa* (Table 4.3).

Chapter 4 The Significance of Water for the Structure and Properties of Collagen and Leather

Table 4.3 Relationship between the T_s and the medium employed for measurement

Composition of the solution employed in T_s measurement	T_s
Water	Reference (57 ℃ ~ 62 ℃)
80% water, 20% ethanol	Δ − 6 ℃
50% water, 50% ethanol	Δ + 6 ℃
100% ethanol	Δ + 15 ℃
100% methanol	Δ + 20 ℃
100% glycerine	Δ + 15 ℃
0.5 mol/L sodium chloride solution	Δ − 4 ℃
4 mol/L sodium chloride solution	Δ + 8 ℃
3 mol/L formic acid solution	Δ − 6 ℃
29% KCNS solution (hydrotrope)	Δ − 30 ℃
100% formamide (hydrotrope)	Δ − 40 ℃

The comprehensive, in some cases unfortunately contradictory results obtained by numerous authors can be explained broadly as follows: native Type I collagen of the corium owes its hydrothermal stability to its inherent structure, with further stabilization by water. An increase in temperature promotes mobility of the water, thus destabilizing the water envelope (water Types II, III), and leads to shrinkage as a result of the helix-coil transition at a T_s specific to the collagen concerned[120]. Two factors are therefore significant for stabilization: firstly, autostabilization of the collagen, and secondly the supplementary stabilizing effect of the water envelope. It is conceivable that a reduction in the water envelope by the facility for greater convergence of the functional groups of the collagen has the simultaneous effect of promoting autostabilization of the collagen: in other words, the original effect of a water envelope reduction, that of reducing the stability, is overcompensated for. The dependency of the T_s upon the concentration of added solvents and electrolytes which is associated with the reversal of effect would in fact be explained by this mechanism. All effects of the various additives and their quantities, which raise or lower the T_s can be attributed to whether they enhance or attenuate one or both of these structure-stabilizing factors. The resulting (net) overall effect as described must be considered.

4.7 The Role of Water in the Conversion of Collagen to Leather

Water, as an integral component of the collagen, is naturally of key importance in the latter's transformation to leather. Sufficient evidence of this is provided by the fact that freshly harvested (native) hide contributes 2.5 to 3 times the collagen's weight in water to the process, a tenth of which is still contained in the finished leather. Between these two limit values, however, are alternating peaks and troughs of water content, which also reflect the role and dynamic nature of the water. A second factor is also at play however: whereas all previous sections have dealt with the "integral water", "external" water now enters the equation which, in the form of process, rinsing and washing water, and acting as an energy carrier and coolant, drastically and unfortunately increases the volume of water consumed during the leather manufacturing process.

The water content changes during the discrete stages of transformation from collagen to leather, owing to the various treatment processes and influences. In accordance with the objective of the study, figures for the water content are given in terms of the respective collagen content. They show that up to and including the beamhouse stage, the changes in the water content affect only the bulk water (Type III water). Not until the tanning stage are water Types II affected by material, mechanical and/or energy influences.

In some process stages, the reduction of the water content is compensated for to some extent by the binding of other substances, such as curing salts, tanning materials, fatliquoring agents, etc. It is therefore by no means safe to compare water content and mass values quoted in terms of collagen. Vegetable-tanned heavy leather is a particularly illustrative example: the mass of the finished leather in this case may exceed 80% of the pelt mass, as the water displaced by tanning and vaporized by drying has largely been replaced by the vegetable tanning material. The same principle applies to all processes in which substances are inserted into the collagen (curing, tanning, wet finishing)[121].

Changes to the water content of hide, pelt, semi-finished product or finished leather are always associated with changes to the volume of the collagen and/or the leather matrix. They are manifested, with the exception of the liming process, as minor changes in thickness, but with changes in surface area as a function of the fiber orientation which may therefore be considerable depending upon the axis (Δ longitudinal/lateral), and therefore also in the leather yield (cm^3 product/100 g collagen). These dimensional changes are subject to a wide range of influences, and likewise vary according to hide type, process control and desired leather type and to the given process factors. Only approximate values can therefore be given.

4.8 Tanning as Dehydration: A Chapter of Tanning Theory

Dr. G. Reich presented a comprehensive study on the subject of tanning theory[1], in which tanning is described here as *dehydratation*. Instead the question addressed here is how tanning materials react at all with the collagen, given that the latter's structural elements all exhibit an integral water coating. Do site interchange and displacement processes occur; is the situation one of competition for facilitated hydrogen bond formation? How should we then imagine the structure of the collagen-tanning material-bound water system? Does mutual penetration take place to form a mixed matrix (suprastructure), or does the water bridge simply migrate from the

Chapter 4   The Significance of Water for the Structure and Properties of Collagen and Leather

collagen to the tanning material? In order for answers to be found to these and other questions, certain issues must first be resolved. The first concerns the relationship between the type and scale of tanning material binding on the one hand and the conceivable displacement of water on the other. Preliminary discussions have already shown that the natural or synthetic tanning materials bound in considerable quantity by hydrogen bonds are of particular relevance to water[64]. For example, during mimosa tanning, water was released from the pelt in approximately the same quantity as that of tanning material absorbed, i. e. the net mass remained the same. This study, also interesting with regard to the methods employed, demonstrated that in the case of vegetable tanning, the pelt volume did not change, i. e. that the volumes of the tanning material taken up and water displaced were the same.

Aldehydes and other tanning materials bound covalently to lysine may be assumed to reduce the volume of water released owing to the nature of their binding and the comparatively low bound quantity[122]. The chrome tanning materials are an intermediate case; it must however be remembered that chrome tanning materials contribute a water "mantle" of their own in the form of their aquoligands and possibly also coordinatively bound water layers.

The water release dealt with so far is therefore on the one hand a consequence of the mass transfer: tanning material for water. To that extent, it is dependent upon the quantity of tanning material taken up. A second mechanism is, however, also at work, namely that tanning gives rise to stiffening of the fibrils and fibers. This reduction in flexibility has the effect of the pelt water being "squeezed out", and is also enhanced by the drum movement. Any tanner in the field knows that water can be squeezed out of a tanned pelt by bending, something which cannot be achieved with the original pelt. In addition, residual floats are observed at the end of non-float processes, and an extension of float length in tannages with normal floats. However, the effects described here all relate solely to changes in the *bulk water*, i. e. the Type III water. They say nothing about the Type II and I water bound to the collagen, and the questions posed at the outset remain unanswered.

The reader's attention is therefore drawn to an earlier study of the water balance of vegetable-synthetic-tanned leather. This study demonstrated that for a given tanning material type, the maximum water vapor sorption, the criterion for binding of Type II water, falls on the one hand with rising tanning index, whilst on the other remaining constant in terms of the collagen content, independent of the tanning index. Differences exist however between tanning materials which reveal something of the bond type. Syntan-tanned leathers, for example, take up less water vapor (blocking of the amino groups by sulphonic acid groups), which is consistent with the known tendency of syntan-tanned leather to dry out (Table 4.4).

Fundamental Collagen Chemistry in Leather Making

Table 4.4 Maximum water sorption of leather subject to different vegetable-synthetic tannages

Tanning material	Tanning index (bound tanning material per 100 g of collagen)	Max. water vapor sorption in %	
		In terms of leather meal	In terms of hide substance
Spruce	18	55	65
	28	50	65
	39	47	65
	59	41	65
Oak	20	52	63
	37	50	68
	59	40	64
	67	35	57
Pellutan EZE	18	47	55
	28	45	58
	39	39	54
	47	38	56
Wofagan	14	51	58
	24	44	54
	31	41	56
	41	37	52

Mention is also made of the finding in the study cited that the tanning agents themselves possess considerable capacity for sorption water vapor, and this capacity differs between the tanning materials and the non-tanning ones. Mimosa extract for example takes up 145% water in terms of dry substance from water vapor-saturated ambient air. 60% of this is accounted for by the tanning material, 40% by the non-tanning materials. The ratio is different for sulphited quebracho, namely 153% total take-up in terms of dry substance, almost 100% of this however by the non-tanning materials. The division into tanning materials and non-tanning materials is achieved by filter analysis. The water vapor sorption of the dried original extract and of the vaporized non-tanning materials in the filtrate of the filter analysis was measured directly, that of the tanning materials calculated from the difference between the two values.

Finished leathers from the widest possible variety of manufacturing processes almost always exhibit water contents of between 12% and 20% when dried in a controlled climate. This apparently wide scatter is narrowed decisively, however, when the water content concerned is calculated in terms of the collagen/hide substance content of the leather. This has also been demonstrated by comprehensive tests on dimensional stability. There is much to suggest that tanning (and also filling and fatliquoring) does not cause collagen to lose its ability to adsorb water, and that in fact the collagen retains its water coating, characterized by Types I and II

Chapter 4 The Significance of Water for the Structure and Properties of Collagen and Leather

water, relatively unimpaired.

In the author's view, these considerations mean that the questions formulated at the beginning of this section regarding the actual structure of the leather matrix in terms of collagen/water/ingredients have not yet been answered with adequate scientific precision. The question could, for example, be examined of whether polymers, resin tanning materials, kaolin and similar fillers, and also fatliquors, do not in fact fill only capillaries and pores, i. e. leaving the water coating of the collagen relatively unchanged.

More exact knowledge of the changes in the water balance of collagen brought about by tanning (and wet finishing) are not only desirable, but also necessary for systematic development of new tanning and retanning materials. The merging of experimental study and computer modelling may be the correct strategy.

Chapter 5 Introduction of Modern Tanning Chemistry

5.1 Vegetable Tanning

Tannins are usually defined as polyphenolic compounds which can precipitate proteins. They are part of a diverse group of polyphenols as secondary metabolites in plants that were originally widely-used in the tanning of animal hides in leather industry[123-125]. Many plant materials contain polyphenols which can be used in tanning. To be effective, their molecular mass must be 500~3000; lower molecular mass fractions in the tannin are referred to as non-tans and higher molecular mass species are gums. Tanning products may be powdered plant parts or aqueous extracts of those parts; the properties they confer to the leather are as varied as the many sources from which they are obtained. The varied chemical structures and stability evidenced among the tannin group can result in their classification as follows: hydrolysable or pyrogallol tannins, subclassified as gallotannins (Chinese gallotannin or tannic acid, sumac, tara, etc.) or ellagitannins (myrobalan, chestnut, oak, etc.), and condensed tannins (mimosa, quebracho, gambier, etc.).

Hydrolysable tannins are sugar derivatives, based on glucose, but may be larger polysaccharides; plant extracts may contain tetrasaccharides as less useful gums[126, 127]. Gallotannins are characterized by glucose esterified by gallic acid as shown in Figure 5.1; esterification may occur directly with the glucose ring or as depside esterification of bound gallic acid. Ellagitannins have sugar cores, esterified not only with gallic acid, but also with ellagic acid and chebulic acid. Examples of structures of hydrolysable tannins are chebulinic acid from myrobalan, chebulagic acid and vescalagin and castalagin of chestnut. Hydrolysable tannins typically raise the T_s of collagen to 75 ℃ ~ 80 ℃. The traditional way to tan hides or skins with vegetable tannins is in pits, where the slow penetration of large reactive molecules can take place over a prolonged period of time. When leather quality was overseen by the Guilds, hide had to stay in the pits for a year and a day, a form of quality assurance. Currently, the tanning period has been remarkably reduced to a few weeks. A feature of pit tanning with hydrolysable tannins is that they deposit "bloom"; natural fermentation breaks down the tannin into sugar acids and precipitates components, such as ellagic acid. The bloom has a filling effect within the collagen fiber structure. This is useful because sole leather is sold by weight, and the organic acid salts provide a buffer against the detanning effects of sulfur oxides and nitrogen oxides in the atmosphere. This latter reaction is known as "red rot", familiar as the cracking and disintegration of bookbinding leathers.

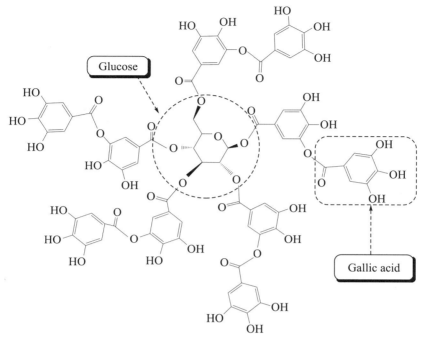

Figure 5.1 Schematic diagram of molecular structure of Gallotannins

Condensed tannins are based on the flavonoid ring system[128-130], shown in Figure 5.2. The A ring usually contains phenolic hydroxyl groups, and the presence of the C ring makes both rings reactive to forming carbon-carbon bonds. The B ring does not exhibit the same reactivity, which often contains the catechol group. So the alternative name for this group of compounds. The condensed tannins are illustrated in the generalized structure and the monomeric units of mimosa and quebracho tannins. The condensed tannins do not undergo hydrolysis, instead they may deposit a precipitate, an aggregate of polyphenol molecules, called "reds" or phlobaphenes. Unlike hydrolysable tannins, which are relatively lightfast, the condensed tannins redden markedly upon exposure to light; this is understandable in terms of their linked ring structure and ability to undergo oxidative crosslinking. Condensed tannins typically raise the T_s of collagen to 80 ℃ ~ 85 ℃, which is higher than that of hydrolysable tannins.

Fundamental Collagen Chemistry in Leather Making

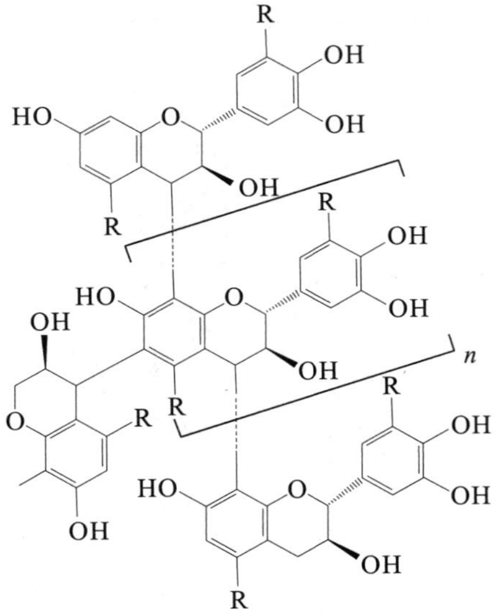

Figure 5.2 Schematic diagram of molecular structure of condensed tannins

Vegetable tannins can react with collagen primarily through hydrogen bonding[131], as indicated in Figure 5.3. This type of interaction is inferred from studies of reaction with polyamides. It is known that polyphenols can also fix to amino sidechains via electrostatic effects with carboxylate or hydrogen bonding with carboxylic acid groups (depending on the pH value). It is also accepted that condensed tannins have an additional mechanism for reaction. This can be related to the effect that they are more resistant to removal by hydrogen bond breakers. For example, treating leather tanned with myrobalan (hydrolysable tannin) with 8 mol/L urea removed 80% of the bound tannin, reducing the T_s by 10 ℃ ~ 20 ℃; whereas, mimosa (condensed tannin) tanned one treated under the same conditions lost only 50% tannin and the T_s fell by 4 ℃ ~ 5 ℃. It has been suggested that this additional interaction can be covalent reaction between the protein and aromatic carbon in the tannin molecules via quinoid structures. Note that the quinone itself can tan protein effectively, which can raise the T_s to 90 ℃ ~ 95 ℃.

Chapter 5 Introduction of Modern Tanning Chemistry

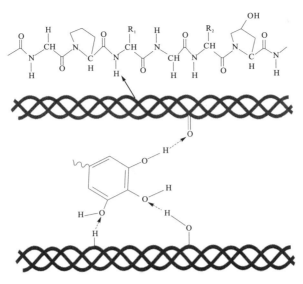

Figure 5.3 Schematic diagram of hydrogen bonding between plant polyphenols and collagen

Modern vegetable tanning, especially for sole leather, is still conducted in the pits. The procedure retains some of the traditional elements. For example, high affinity of polyphenols for protein means that the tannage must start in weak liquors, and the hides are progressively moved through a series of pits containing increasing concentrations of the vegetable tannins. Depleted liquors are strengthened by topping up with stronger liquors, and in this way the hides go one way via the system. The liquors can go in the opposite direction, e. g. the countercurrent method of pit tanning. The use of extracts, rather than the plant material itself, allows highly concentrated solutions to be employed and by warming the pits, "hot pitting", and therefore the whole process takes only a few weeks. Vegetable tanning can also be conducted in rotating drums, but there is a need to make the hide less reactive, to allow the tannins to penetrate into the thick hide. The oldest method is to pretan with synthetic analogues, syntans, and one of the newer methods is to precondition the hide with polyphosphate, which has a weak tanning action. In this way, vegetable tanning can be shortened to a few days.

The study of plant polyphenols is an active field, not only because of application to tanning technology, but also because it is a fruitful area for a wide range of products, either isolated compounds or chemically modified polyphenols. That range includes adhesives (especially wood adhesives), fisheries, beverages manufacturing, animal feed, biosourced foams, wood preservatives, corrosion inhibitors, polyurethane surface coatings, epoxy adhesives, binders for Teflon coatings, and etc[132].

5.2 Mineral Tanning

A review across the Periodic Table of the tanning effects of simple inorganic compounds reveals that many elements are capable of being used to make leather. However, considering their practical criteria of effectiveness, availability, toxicity and cost, the number of useful options is much reduced. In all cases, the benchmark for comparison is tannage with Cr(Ⅲ). $T_s > 100$ ℃ is easily achieved, and it is readily available with large reserves in Southern Africa. Moreover, it is relatively cheap and has minimum health hazards or environmental impact.

5.2.1 Chromium(Ⅲ) Salts

There is a fortuitous coincidence of reactivity in chrome tanning. The reaction occurs at ionized carboxyl groups; and aspartic and glutamic acid sidechain carboxyl groups have pK_a values 3.8 and 4.2, respectively, providing a reaction range at pH 2 ~ 6. Cr(Ⅲ) forms basic salts in the range pH 2 ~ 5, although the useful range is pH 2.7 ~ 4.2 in practice, where the basicity ranges from 33% to 67%. Schorlemmer basicity is defined by the number of hydroxyl groups associated with the metal ion, relative to the maximum number allowed by the valency. In that useful range, the number of chromium atoms in the molecular ion increases from 2 ~ 3 to > 3, and the availability of ionized collagen carboxys increases from 6% to 47% of the total number.

Hide is prepared for tanning by pickling with sulfuric acid in a solution of sodium chloride, the neutral electrolyte is necessary to prevent osmotic swelling of the collagen. In conventional tanning process, chrome tanning is initiated at pH 2.5 ~ 3.0 by using 33% basic Cr(Ⅲ) sulfate in the form of spray dried powder, obtained from sulfur dioxide reduced chromic acid. During the tanning process, the pH is raised to 3.5 ~ 4.0, causing the number of reaction sites on the collagen to increase and the chrome species to increase in size. Starting the process under conditions of low reactivity of both collagen and chrome favors fast penetration of chrome into the substrate, but slow reaction. Increasing the pH increases the reactivity of both components of the reaction, resulting in reduced penetration rate. To obtain a continuing balance between reaction rate and penetration rate is part of the tanner's art, this is not simple.

The changes that occur in chrome species are set out in Figure 5.4. The formation of olated species was first used to explain the hysteresis in delayed back titration of basified Cr(Ⅲ) and the compound containing both hydroxyl and sulfate bridges is well characterized. Further change into the oxolated (oxo bridged) species is postulated to happen during ageing after tanning.

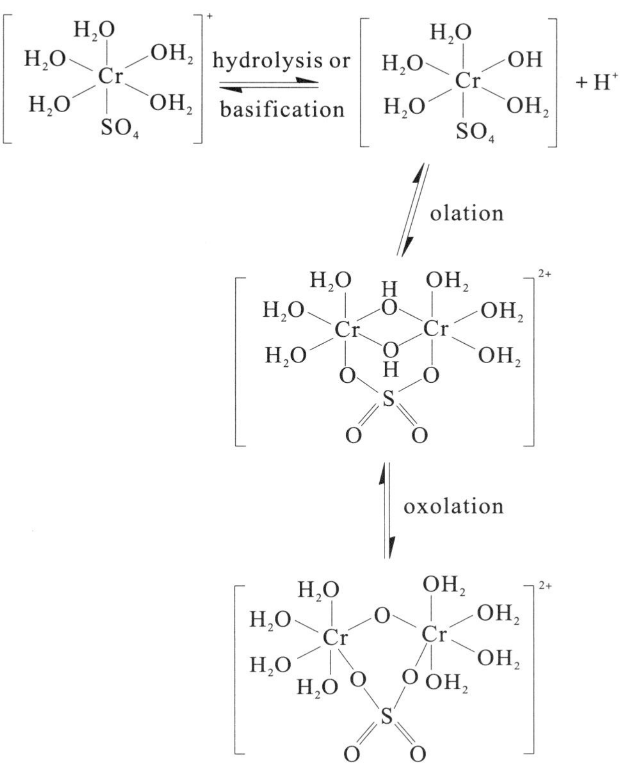

Chapter 5 Introduction of Modern Tanning Chemistry

$$\left[\begin{array}{c} \text{H}_2\text{O} \\ \text{H}_2\text{O}-\overset{|}{\underset{|}{\text{Cr}}}-\text{OH}_2 \\ \text{H}_2\text{O} \quad \text{OH}_2 \\ \text{SO}_4 \end{array}\right]^+ \underset{\text{basification}}{\overset{\text{hydrolysis or}}{\rightleftharpoons}} \left[\begin{array}{c} \text{H}_2\text{O} \\ \text{H}_2\text{O}-\overset{|}{\underset{|}{\text{Cr}}}-\text{OH} \\ \text{H}_2\text{O} \quad \text{OH}_2 \\ \text{SO}_4 \end{array}\right] + \text{H}^+$$

$\Updownarrow$ olation

(dichromium species with bridging OH and SO₄, charge 2+)

$\Updownarrow$ oxolation

(dichromium species with bridging O and SO₄, charge 2+)

Figure 5.4 Changes in chromium (Ⅲ) species with pH value

The presence of coordinated sulfate is necessary for the efficient reaction of Cr(Ⅲ). Selective tannage with the isolated dichromium species is 11 ℃ lower if sulfate is not present and 15 ℃ lower for tannage with the trimeric chromium species. The coordination of ligands to Cr(Ⅲ), to modify the properties of the salt, is routinely exploited as "masking". Monodentate ligands, especially formate, may be applied at different ligand to metal ratios, depending upon the degree of effect desired. Reduction of cationic charge and statistical reduction in the number of reaction sites on the molecular ion make the species less reactive to the collagen, hence enhancing penetration rate. In addition, such masking can increase the pH value at which the salt precipitates; in these circumstances, the final pH of the tannage may be elevated beyond that of unmasked tannage, thereby enhancing the reactivity of the collagen. In this way, the reaction rate can be accelerated, but without the same effect of increasing the size of the chrome species. Masking with bidentate ligands, which are capable of crosslinking chromium ions, can cause a big increase in size, resulting in a statistically higher number of reaction sites per molecular ion. Dicarboxylates containing two or more methylene groups perform this function, but phthalate is the salt of choice. Whichever masking salts are used, they are usually added to the tanning bath, and the masking reaction proceeds at the same rate as the tanning reaction. This is because the reactions are identical and the formation of carboxyl complexes with Cr(Ⅲ)[133].

The origin of the high hydrothermal stability of Cr(Ⅲ) tanned leather is interesting[134]. By selectively deactivating collagen carboxyl groups by esterification and amino groups by acylation, it is found that hydrothermal stability can be controlled by reaction between Cr(III) and the carboxyl groups, although the removal of these reaction sites does not result in zero chrome fixation. Hence, chrome fixation can occur in three ways:

· Covalent reaction between one chromium ion and one carboxyl, this is unipoint fixation.

· Covalent crosslinking between one chromium ion and at least two carboxys, this is multipoint fixation.

· Hydrogen bonding between chromium species and the protein, especially along the polypeptide backbone.

It has always been assumed that the reaction determining the high hydrothermal stability is multipoint fixation; unipoint fixation probably provides little hydrothermal stability. There are many examples of hydrogen bonding in tanning, including vegetable tanning, which confer only moderate hydrothermal stability.

The effect of Cr(III) on crosslinking can be calculated as follows. $T_s > 100$ °C is achieved with a chrome content of approx. 2.5% Cr on dry leather mass. Since complexes are binuclear or bigger, 2.5% Cr = 0.5 g atom kg^{-1} = 0.25 mol molecular ion kg^{-1}. Note that 1 kg of dry collagen contains 1 mol of carboxyl groups, and therefore only one quarter of the carboxys react with chrome. But it has been estimated that only 10% of the bound chrome is involved in crosslinking. Therefore, only 1/40 of the carboxyl groups are involved in crosslinks with chrome. Collagen contains 11.6% acidic residues (Asp + Glu) and, since there are 1052 residues per chain, this is equivalent to 120 residues per chain = 360 residues per triple helix. Therefore, the number of residues reacted in crosslinks = 9. Or, the number of crosslinks per triple helix = 5. The size of the cooperating unit in the chrome tanned leather is calculated to be 206 residues, so that suggests the effect of a chromium crosslink extends $ca.$ 50 residues either side of each end of the crosslink, which is almost two complete twists of the triple helix.

The options for crosslinking are threefold: they are intra single helix, intra triple helix, and inter triple helix. From the known amino acid sequences of the α1(I) and α2 chains of Type I collagen, the relative numbers of the crosslink types can be calculated. It is assumed that aspartic acid and glutamic acid sidechain carboxyl groups are equally reactive, and a reaction can occur between two sidechains no more than three residues apart. Note, hydrolysis of amide groups on the sidechains of Asp and Glu does occur during the alkali treatment of hide, typically to the extent of $ca.$ 50% converted to carboxyl groups. The assumption of equal reactivity may not be justified, since model studies indicate that poly glutamic acid is less reactive to Al(III) than poly aspartic acid, presumably entropy controlled.

It is interesting to speculate that, if the most likely stabilizing crosslinks are between triple helices, then 50% deamidated collagen could form 52 crosslinks. But if only 10% of bound chrome is involved in crosslinking, perhaps only 10% of possible crosslinks are formed; this is in agreement with the calculation presented above. Experiments on native and variously deamidated collagens do indicate a difference in tanning effects, but it is not clear whether this is due to a difference in the availability of reaction sites or to a difference in the distribution of potential crosslinking sites.

Chrome fixation is accelerated by elevated temperature and pH value; the higher the chrome content of the leather, the higher is the T_s. But, the industrial requirement is to obtain high shrinkage from the minimum amount of chrome used. Studies have shown that temperature and pH value do not have equivalent effects, in which tanning effectiveness is measured by the rise in T_s per unit of bound chrome. It can be seen that under tanning condition of constant temperature and pH value, rise in T_s is controlled by pH value. Furthermore, tanning effectiveness is better when low basicity chrome salts are basified during tanning, than if moderate or high basicity salts are employed at the beginning of tanning.

Closer examination of the results for tanning at 25 ℃ and pH 3.5 reveals that there is a maximum effectiveness of tanning. Optimum tanning occurs at an offer of 0.7% Cr_2O_3, although this corresponds to a T_s of only 77 ℃. Here, 4% Cr_2O_3 in the leather corresponds to a T_s of 107 ℃; this chrome content is the industry standard to provide a degree of guarantee that the leather will withstand boiling water for at least 2 min.

Chrome tanned leather is highly versatile, largely due to the low level of tanning agent needed to achieve the desired stability. This means that the variety of retanning materials which might be applied to the part processed leather can produce a wide range of final products. Indeed, from any one chrome tanned cattle hide, it is possible to produce a sole leather or combat boot upper leather or softer shoe upper leather or upholstery leather or garment leather, all as full grain or suede leathers.

The part processed chrome tanned leather is called "wet-blue", because it is wet and it is blue. The color comes from the protein carboxyl complexes of Cr(Ⅲ). The color also depends on whatever masking salts are used, typically it is bright pale blue, but it can range from pale green to purple. This does introduce some difficulties regarding dyeing, but it has not restricted the color range of fashion leathers to any great extent, with the possible exception of pure pastel shades.

5.2.2 Aluminium(Ⅲ) Salts

The use of potash alum in leather making is almost as old as leather making itself. It is known that the Egyptians used it 4000 years ago. Throughout tanning history, alum was often used in conjunction with vegetable tannins. For example, in medieval times, Cordovan leather (from Cordoba in Spain, hence the name cordwainer, meaning shoemaker) was in widespread use in Europe, made by vegetable tanning then dyeing with cochineal.

Used by itself, alum (solution pH 2) interacts only weakly with collagen, scarcely raising the T_s and therefore having little leathering effect. However, in a mixture of water, salt, flour (to mask the aluminium ion and fill the fiber structure) and egg yolk (the lecithin content is an effective lubricant), skin can be turned into a soft, white, leathery product, traditionally used in the past for gloving. But, even in this case, the T_s is not raised. Therefore, it is possible to discriminate between leathering and tanning, and the aluminium salt can be washed out of the leather if it gets wet; for these reasons, this process is called "tawing", to distinguish it from tanning[135].

The reaction sites for Al(III) are the collagen carboxyl, but unlike Cr(III) to which it bears a superficial resemblance in a tanning context, Al(III) does not form defined basic species nor does it form stable covalent complexes with carboxyl groups. The interaction is predominantly electrovalent, accounting for the ease of hydrolysis, which can be optimized for tanning by modifying aluminium sulfate with masking salts, such as formate or citrate, and basifying the tannage to pH 4, close to the precipitation point. There is a rule of thumb in tanning technology, that any metal salt has its greatest tanning effect just before it precipitates. In this way, reversibility of tannage is minimized, and T_s as high as 90 ℃ can be achieved. Basic Al(III) chlorides are also well known in leather making, and several commercial tanning formulations are available. As solo tanning agents, they are slightly superior to salts based on the sulfate. However, the leathering effect of Al(III) is inadequate, producing firm leather, which may dry translucent due to the fiber structure resticking. Therefore, as tanning agents, Al(III) salts have limited value in leather tanning.

Where aluminium salts are useful is their ability to accelerate chrome tanning. Following a pretreatment with Al(III), the incoming chrome displaces it. It is known that the rate of exchange of solvate ligands is 10^6 times faster at Al(III) than at Cr(III). It is possible to postulate that Al(III) can react quickly with other ligands, including collagen carboxyl, in a loose association that is entropy favored or not disfavored. When Cr(III) enters the system, the activation barrier is lower. This is because the reactants are effectively shifted along the reaction coordinate and the reaction can be accelerated. Since Al(III) is present only at low concentration and plays a facilitating role, this process appears to be catalytic.

5.2.3 Titanium(IV) Salts

In tanning terms, the chemistry of Ti(IV) salts lies somewhere between Al(III) and Cr(III). Empirically, the chemistry is dominated by the titanyl ion TiO^{2+}, but the species are chains of titanium ions bridged by hydroxyl and sulfate ligands, like Cr(III). However, the coordinating power is weak with respect to carboxyl complexation, so the interaction is more electrostatic than covalent. The traditional use for Ti(IV) in tanning was in the form of potassium titanyl oxalate, to retan vegetable tanned leather for hatbanding, a product for which demand has reduced in the latter half of the 20th century[136]. Solo titanium tanning is only moderately effective, because large quantities are required to achieve the highest $T_s > 95$ ℃, but this causes the leather to be overfilled, although remaining soft. In addition, high hydrothermal stability is only achieved when the collagen is pretreated with phthalate.

An advantage of tanning with Ti(IV) is that it is a colorless tannage and therefore makes white leather[136, 137]. Hence, it has found application in tanning sheepskins with the wool on. This is a problem area, firstly because the use of Cr(III) produces a discoloration by reaction with the

Chapter 5　Introduction of Modern Tanning Chemistry

partially degraded keratin at the weathered wool tips and secondly because the reaction must be conducted using high solution to skin ratios, to avoid tangling the wool. However, one of the problems of using Ti(Ⅳ) salts is their tendency to hydrolyze and precipitate in dilute solution.

It was argued that a mixture of salts might produce a tanning complex of better value than the individual salts[138,139]. It was found that mixtures of the metal sulfates could be stabilized against hydrolysis at pH 4 by complexing (masking) with gluconate $[HOCH_2(CHOH)_4CO_2^-]$ and in this way the mixed salt could be used to tan white sheepskin rugs; in more concentrated tanning solutions, T_s as high as 95 ℃ can be achieved. It is known that Al(Ⅲ) and Ti(Ⅳ) salts can form mixed complexes in which the ions are bridged by hydroxy and sulfate ligands, but they still interact with collagen in a primarily electrostatic manner.

5.2.4　Zirconium(Ⅳ) Salts

The development of zirconium tannage is relatively recent, but it soon gained industrial acceptance. Zr(Ⅳ) might be expected to display similar tanning properties to Ti(Ⅳ). It can be seen that the tanning power exceeds that of Al(Ⅲ), but in no way matches Cr(Ⅲ). Whilst the tanning effects of Zr(Ⅳ) and Ti(Ⅳ) are similar, the chemistries of their salts are different. Zr(Ⅳ) salts are characterized by eight coordination and high affinity for oxygen, resulting in a tetrameric core structure. The basic unit of structure is four Zr ions at the corners of a square, linked by diol bridges, above and below the plane of the square.

By hydrolysis or basification, the tetrameric units can polymerize by forming more diol or sulfate bridges. In this way, zirconium species may be cationic, neutral or anionic and large ions can form. So, tanning may involve all the polar sidechains of collagen, those bearing carboxyl, amino or hydroxyl groups. Hydrogen bonding via the hydroxyl groups in the Zr(Ⅳ) species is an important feature of the tanning reaction; together with the filling effect by the big molecules, the overall tanning effect is somewhat similar to tanning with plant polyphenols. Therefore, zirconium tanning has been referred to as the inorganic equivalent of vegetable tanning[140,141].

Zr(Ⅳ) is not often used as a solo tannage. This is partly because of its indifferent effectiveness at raising the T_s, and partly because the acidity of the salts and the vulnerability to hydrolysis mean that they must be applied at high concentration and at pH < 1, therefore running the risk of osmotic swelling in the hide. Hence, its main use is for retanning, to fill and firm the grain or to make better suede.

5.3　Oil Tanning

The familiar wash leather (chamois or chammy) is tanned with unsaturated oil[142-145]. The preferred agent is cod liver oil. Useful oils contain fatty acids, either free or as glycerides, which are polyunsaturated. The degree of unsaturation is critical. This can be owing to the fact that if there is too little unsaturation, the oil will not oxidize readily, and therefore function only as a

Fundamental Collagen Chemistry in Leather Making

lubricant; if there is too much unsaturation, the oil will crosslink itself and harden with oxidation, like linseed oil.

It is thought that oil tannage may be due in part to an aldehyde reaction and to polymerization of the oil; the presence of the latter effect could account for the difference between the characteristics of oil and aldehyde tanned leathers. The situation is further complicated by the observation that oil tanning hardly raises the T_s of collagen; so, this is a leathering process rather than a tanning process, based on the accepted criteria of tanning.

Oil tanned leather exhibits the interesting Ewald effect: if the leather is heat shrunk in water at 70 ℃, but immediately placed in cold water, it rapidly relaxes to regain $ca.$ 90% of its original area and this is repeatable. Normally, heat shrinking is an irreversible phenomenon. Furthermore, if the wet leather is held under tension whilst it is being heat shrunk, the dried leather remains soft and flexible, unlike other leathers which may come hard and brittle. This process, known as "tucking", is used to mould leather to a desired shape, but keeping its feel.

The most remarkable feature of oil tanned leather is its hydrophilicity, surprising considering its tanning process. A well-tanned chamois leather is expected to take up at least 600% water on its dry mass and to be hand wrung to 180%. Also, this must be repeatable after drying. In use, no grease must be exuded to cause smearing. A synthetic version of oil tanning is to use a sulfonyl chloride, which reacts predominantly with the amino groups on collagen. Clearly this is not a crosslinking reaction, so it is not surprising that the T_s is not raised by this tannage; but there is a powerful leathering effect and the product exhibits similar properties to oil tanned leather.

It was pointed out that acrolein is produced during the oil tanning reaction. Indeed, acrolein itself can be used to make a leather similar to oil tanned. But, although it is not used itself for this purpose (for toxicity reasons), it is used indirectly, as a component of wood smoke. The traditional method of preserving hides and skins used by the plain dwellers, such as the North American Indians and the Mongols, is to use brains tanning. In this process, the animal brain is partly cooked in water, so it can be mashed into a paste, which can be worked into the pelt. The leathering effect turns the skin into a soft, open structured leather, buckskin, largely due to the lubricating power of the phospholipids of the brain. The Sioux Indians have a saying: *"Every animal has enough brains to tan its own hide."* The leathering effect is serviceable, as long as the pelt is not rewetted, because then it will harden on drying due to the fibers resticking. To make the leather resistant to wetting, the solution is to smoke it over a wood fire; the multiplicity of free radical and other reactions do not adversely affect the handling qualities and they are made permanent.

5.4 Aldehyde Tanning

5.4.1 Formaldehyde Tanning

The archetypal aldehyde tannage is with formaldehyde, probably most familiar in preserving biological specimens or in embalming. Reaction occurs primarily at amino groups. The N-hydroxymethyl group is highly reactive and crosslinking can occur at a second amino group. In this way, the T_s can typically be raised to 80 ℃ ~ 85 ℃. However, the crosslinking is relatively inefficient, probably because the formaldehyde species are not monomeric. Amongst the species formed in solution is paraformaldehyde, $HOCH_2(CHOH)_n CHO$. The presence of polyhydroxy species and their reaction with skin produces a white, spongy and hydrophilic leather, although the absorptive property of oil tanned leather is not matched.

The health and safety implications associated with formaldehyde mean that its use as a tanning agent is effectively banned. The only remaining common functions are for fixing casein, which is fused to the grain surface in glazing operations, or to impose permanent straightening to the wool of sheepskin clothing leathers, by reacting with the keratin under conditions of wet heat and tension.

5.4.2 Glutaraldehyde Tanning

Of the many mono- and multi-functional aldehydes which might be used for tanning (and all can be made to work), only glutaraldehyde and its derivatives have found commercial acceptance, with the possible exception of the more expensive starch diadehydes. Glutaraldehyde can also be used as a crosslinking agent for collagen-based biomaterials[146]. The reactions of glutaraldehyde are set out in Figure 5.5. The crosslinking options are wider than for simple aldehydes, but the result is the same, a T_s of 85 ℃ maximum. In the same way that formaldehyde is not a simple species in solution, glutaraldehyde is polymerized, shown in Figure 5.6. The terminal hydroxy groups of the polymer are active and capable of reacting with amino groups[147]. The polymer itself can interact with the collagen peptide links by hydrogen bonding via the alicyclic oxygens and so the leather is given its spongy, hydrophilic character.

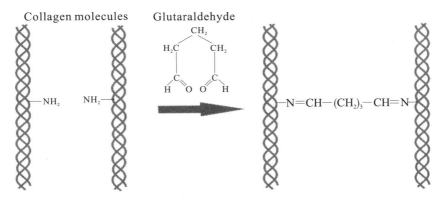

Figure 5.5 The tanning reactions of glutaraldehyde

Fundamental Collagen Chemistry in Leather Making

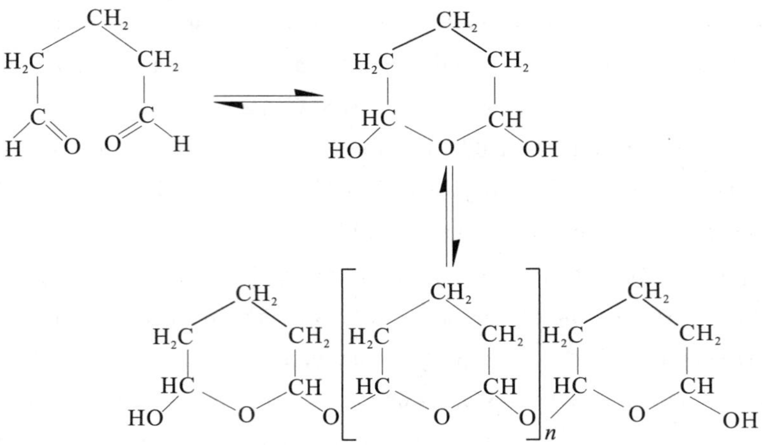

Figure 5.6 The reactivity of glutaraldehyde

Tanning with glutaraldehyde itself confers a yellow-orange color to the leather, which is undesirable. Several attempts have been made to modify the chemistry, to prevent color development, including making the monobisulfite addition compound or hemiacetals, but none has been totally successful. Glutaraldehyde is coming under scrutiny with regard to health and safety implications, so it too may have to be phased out of the tanner's options.

5.4.3 Oxazolidine Tanning

An alternative to aldehyde tanning, but which retains the essential reactions, is to use oxazolidines, developed less than 20 years ago. These compounds are alicyclic derivatives of an amino alcohol and formaldehyde; under hydrolytic conditions, the rings can open, to form an N-hydroxymethyl compound, which can react with one or more amino sites in an effective though acridly odiferous tannage (Figure 5.7).

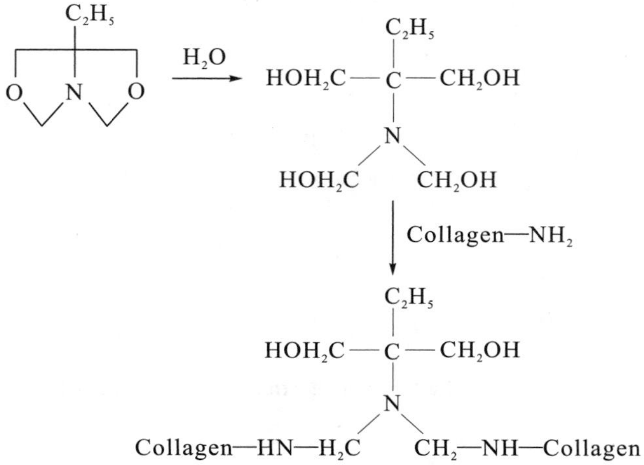

Figure 5.7 The tanning reactions of oxazolidine

Oxazolidines are heterocyclic derivatives obtained by the reaction of amino-hydroxy compounds with aldehydes. Oxazolidine E, which is predicted to cross-link protein molecules through the formation of a carbocationic intermediate formed as a result of ring opening, has been previously shown to impart hydrothermal and mechanical stability to collagen, as manifested by the higher shrinkage temperature of the treated hide[148]. An important effect of these agents is their influence on the chrome tanning reaction, promoting the fixation; it is not known whether this is a function of the chemistry analogous to ethanolamine (see above).

5.4.4 Active Hydroxy Compounds

The reactivity of the carbon in N-hydroxymethyl compounds is not the only example of an active hydroxyl. Tetrakis (hydroxymethyl) phosphonium sulfate/chloride (THPS or THPC) are organophosphine derivatives as typical water-soluble quaternary phosphonium salts. The phosphonium salts are available as bactericides, and can be utilized for a wide variety of industries including wastewater treatment agent[149], and environmentally friendly biocide due to its low environmental toxicity and no potential bio-accumulation[150]. In leather industry, it is highly considered as an effective non-metal tanning agent for wet-white leather manufacture, and the aldehydic reaction mechanism between active hydroxymethyl groups of THP molecules and functional amino groups of collagen is generally accepted (Figure 5.8)[151, 152]. Moreover, it can confer leathers with fine hydrothermal stability and promote the uptake of anionic fatliquoring agents and dyestuffs in post-tanning procedures. Nevertheless, THPS tanning has several shortages such as potential HCHO release[153], poor yellowing resistance and light fastness of the leathers. Moreover, a globally growing awareness of the need to control phosphorus emissions in leather processing, which is reflected in increasingly stringent regulations, has made phosphorus removal more widely employed in wastewater treatment[154]. The combination tannages based on synergistic interactions of THPS with vegetable tannins[155], aluminum salt[156], and clay nanoparticles[157], have attracted considerable interests to acquire desirable physical performances and preferable ecological benefits.

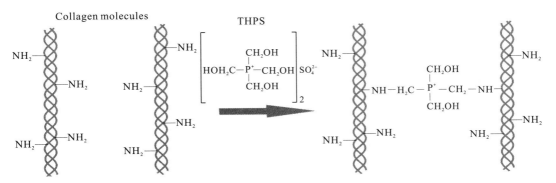

Figure 5.8 The tanning reactions of THP

Fundamental Collagen Chemistry in Leather Making

5.5 Syntans

The term syntan means synthetic tanning agent. This class of tanning agents was introduced early this century, with the purpose of aiding vegetable tanning, although the range of reactivity currently available means that they may serve several different functions. They are classified into three types according to their primary properties.

5.5.1 Auxiliary Syntans

These compounds are frequently based on naphthalene and the base material is sulfonated to a high degree and then may be polymerized by formaldehyde, illustrated in Figure 5.9. The products are usually relatively simple chemical compounds. The presence of the sulfonate groups means that these compounds can interact strongly with the amino sidechains of collagen at pH < 6. In this way, reaction sites for vegetable tannins can be blocked, promoting penetration via the hide cross section. Meanwhile, they can solubilize the aggregated phlobaphenes of condensed tannins, thereby reducing reaction with the hide surfaces. Similarly, they can disperse acid dyes (most commonly used in leather making) and reduce the reactivity of the leather to dyeing, producing more level coloring.

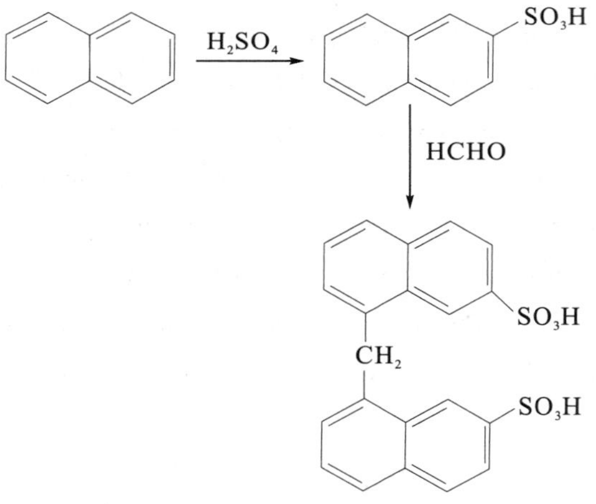

Figure 5.9 The synthesis of syntans

A further function of these simple reagents is to act as non-swelling acids for pickling to low pH, to avoid osmotic swelling. The auxiliary syntans are characterized by their low tanning power. Some may have tanning properties, by virtue of their phenolic hydroxyl content, from the base material. The tanning power is a function of the number of phenolic hydroxyl groups in the molecule, and the degree of swelling they produce in pickling is inversely proportional to the rise in T_s conferred by the "non-swelling acid".

5.5.2 Combination or Retanning Syntan

These syntans are usually based on simple phenolic compounds and are synthesized by the base materials polymerized with formaldehyde and then the product may be partially sulfonated. The products are more complex than the auxiliary syntans, having higher molecular masses, and may be crosslinked in two dimensions. Their enhanced tanning functionality means that they can confer hydrothermal stability and their larger molecular size means that they can have a filling effect. Because they are relatively small polymers, with consequently weak tanning power, these syntans work best as retanning agents. They are applied after main chrome tannage to modify the handling properties of the leather.

5.5.3 Replacement Syntans

By increasing the tanning power of syntans, the agents may be classified as replacement syntans, by which it is meant that they could replace vegetable tannins. These syntans can be used for solo tanning because their properties of tanning are comparable with plant polyphenols. There is no clear distinction between the retanning syntans and the replacement syntans, the difference lies in the degree of the effects. Base materials for syntans can range from the simple to the relatively complicated. In addition, the bridging groups may be more diverse, including dimethyl methylene, ether, urea. They rely less on sulfonate groups for their reactivity, but may incorporate some sulfonic acid functionality during synthesis. The replacement syntans vary in their effects on leather, but can produce properties similar to vegetable tannins, including raising the T_s to 80 ℃ ~ 85 ℃. They are still used to prepare hide to receive vegetable tannins, though they can be used in their own right, to make leather that is more lightfast than vegetable tanned leather; a common use is for making white leather.

5.6 Combination Tanning

In the modern leather industry, the preferred method of tanning is to use Cr(Ⅲ) salts, because they are versatile, effective, easy to apply and have low environmental impact. Currently, chrome tanned leather shares a high proportion of all leather articles worldwide. Some of chemicals used in post-tanning procedures can lead to the possible conversions of Cr(Ⅲ) to Cr(Ⅵ). The formation of Cr(Ⅵ) can cause severe allergic contact dermatitis in human skins and can elicit dermatitis at very low concentrations. In 2014, a more stringent restriction on the Cr(Ⅵ) concentration of leather articles was revised by the European Commission in Annex ⅩⅤⅡ of regulations of Registration, Evaluation, Authorization and Restriction of Chemicals(REACH). In the new restrictions, leather articles, coming into direct and prolonged or repetitive contact with the skin, should not be placed on the market if they contain Cr(Ⅵ) in concentrations equal to or higher than 3 mg/kg. The restriction and risk of Cr(Ⅵ) have directly affected the production, consumption and circulation of chrome tanned leather articles. Despite the evidence to the contrary, the latter aspect of Cr(III) is not wholly accepted by regulatory authorities (with the exception of the USA) and consent limits for discharge of chromium are increasingly stringent. Consequently, although there is a commitment to continuing to use chrome, the industry is constantly searching for a viable alternative.

The ideal tannage to rival chrome tanning should incorporate the following features:
- High hydrothermal stability, $T_s > 105$ ℃.
- No metal salts.
- White or pale colored leather.
- Lightfast.
- Low environmental impact.
- Comparable cost.

The most difficult criterion to achieve is that specifying T_s, which has hitherto been impossible to achieve with organic compounds alone. However, recent developments have indicated that the target is achievable.

5.6.1 Semi Metal Tanning

The only established organic tannage capable of producing leather with high hydrothermal stability is that in which the collagen is first tanned with vegetable tannin, then retanned with a metal salt, preferably Al(III). The semi alum leathers made with condensed tannins typically have T_s ca. 90 ℃ (mimosa is exceptionally higher, possessing pyrogallol groups), whilst the semi alum leathers made with hydrolysable tannins have T_s of 115 ℃ ~ 120 ℃. There is a correlation between T_s and the presence of the pyrogallol group. This effect can be seen even by treating collagen with catechol or pyrogallol themselves, then retanning with Al(III); resulting T_s are 71 ℃ and 98 ℃, respectively, and the same pattern is obtained for more complex polyphenols. The synergistic interaction between the polyphenol and the Al(III) may arise from one of the following options: (1) Collagen—Al—veg—Al—Collagen, (2) Collagen—veg—Al—veg—Collagen, (3) Collagen—veg—Al—Collagen.

It is known that applying the aluminium salt before the vegetable tannin produces only moderate T_s, characteristic of aluminium alone. Therefore, the first and third options are unlikely. The most probable mechanism is for the Al(III) to crosslink the vegetable tannin. In effect, the crosslinking polyphenol on collagen is itself crosslinked, to form a matrix within the collagen matrix, to stabilize the collagen by a multiplicity of connected hydrogen bonds in the new macromolecule.

Because of the presence of pyrogallol groups, semi metal tannages are confined to the hydrolysable tannins and the condensed tannin mimosa. Many metals can perform this function, depending on the affinity for phenolic hydroxy, but Al(III) is the best. It is probable that this tannage is thousands of years old, because of the availability of native potash alum, which must

Chapter 5 Introduction of Modern Tanning Chemistry

have been used together with vegetable tannins, if only for the purposes of mordanting for dyeing.

5.6.2 High Stability Tannages Based on Natural Polyphenols

The alternative to reaction at the phenolic hydroxyl groups is to exploit the reactivity of the A and C rings of the condensed tannins. Aldehydes other than formaldehyde can perform this crosslinking function, although at much slower rates; some of those reactions are important for making wood glues. This type of chemistry has been used in tanning, but the perceived benefit was to reduce the leachability of vegetable tannins from leather.

The reason for using elevated temperature is because the usual control parameter of pH value is not allowed; aldehyde or other active hydroxy tannage is accelerated by pH > 7, but vegetable tannins are stripped from collagen at pH > 6, alternatively, vegetable tannins are firmly fixed at pH < 4, but aldehyde tannage is very slow at pH < 6. Hence, the reaction can only be driven by heat. The most effective and most temperature dependent reaction is crosslinking with oxazolidine. This reaction has been further studied, to determine the magnitude of the synergistic effect.

5.6.3 Synthetic Organic Tanning

The advantage of tannages based on plant polyphenols is that the reagents are obtained from a renewable resource. But the primary disadvantage is the presence of the natural components of the extract that do not contribute to the chemistry of the process. This means that the reaction is not precisely controlled. Clearly, one option would be to isolate the preferred species and that is a possibility for the future, but at an additional cost. A better option might be to target more precisely the required properties of the organic system and to synthesize the reagents.

Recently, a major breakthrough was claimed; it is based on the use of melamine-formaldehyde polymers[158-160], shown in Figure 5.10. It was found that oligomers could be further reacted *in situ* on collagen with additional crosslinker, to raise the T_s to previously unachievable high values. The chemistry is analogous to the crosslinking of vegetable tannins, insofar as the pH requirements for the components are mutually exclusive and the reaction must be driven by elevated temperature. Not all crosslinkers will work equally well in this reaction; for example, collagen tanned with 10% melamine resin then reacted overnight at 50 ℃ with glyoxal or oxazolidine had a T_s of 106 ℃, but crosslinking with the organic phosphonium salt produced 112 ℃.

Fundamental Collagen Chemistry in Leather Making

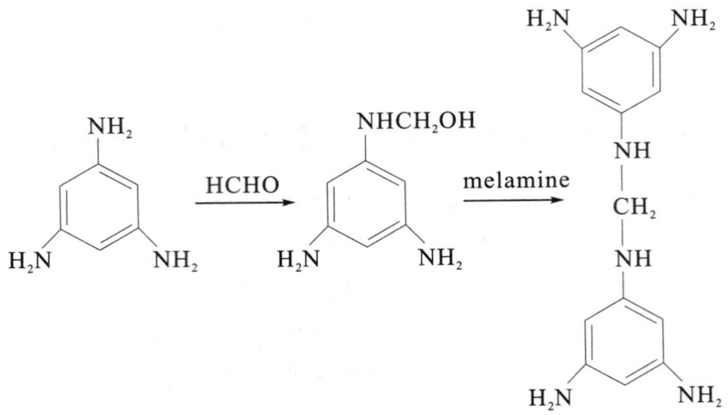

Figure 5.10 Melamine-formaldehyde polymers

Not all melamine resins work, indeed most of the available commercial products yield only moderate T_s, even in the presence of the preferred crosslinker; this is a tanning option that is well known. However, this new tannage depends on a property of melamine resins to aggregate; relatively small oligomers clump together to form larger particles and it is in this form that they usually react with collagen. The hitherto unobserved effect of high temperature processing is to break those aggregates, so that the polymer will react in the form of the oligomers. This is owing to the effect of reaction temperature on polymer particle size and consequent T_s clearly exhibits a preferred particle size range of 50~60 nm.

Furthermore, the particle size effect outweighs even the melamine: formaldehyde ratio, since the relationship between T_s and particle size is the same for melamine to formaldehyde ratios of 5 or 7, typically used in commercial products. These new tannages satisfy the criteria of hydrothermal stability, metal free, white and lightfast. An additional benefit is the speed of the reaction; maximum T_s is reached in 3 h, compared with the 15 h typically required for chrome tanning. The only problem in leather terms is that the leather is weaker than a chrome tanned counterpart. This is due primarily to the filling effect of the resin and the loss of strength is proportional to the amount of resin used. It is probable that the tannage has not been optimized with regard to applying a single or major resin species, so by targeting the reaction more precisely, by minimizing the quantity of resin required to obtain the desired T_s, the strength of the leather will be maximized.

This could be the basis of the tanning chemistry in the future. The remaining questions are: what is the level of the environmental impact of the change in tanning methods and what is the cost in comparison to chrome tanning? The first question cannot be answered until a full environmental impact study is undertaken, but it is known that melamine-formaldehyde polymers are non-toxic and the crosslinkers are safe if handled properly. Cost will depend on the scale of adoption; the basic cost is going to be higher than for Cr(III) salts, but savings may be made through reduced quantities of dyestuffs and lubricants needed for the new leathers.

Biography

1. Reich G. From Collagen to Leather—the Theoretical Background[M]. BASF Service Center Media and Communications: Ludwigshafen, 2007.
2. Brown E. M. Collagen Networks: Nature and Beyond, in Adances in Biopolymers[M]. American Chemical Society, 2006: 36-51.
3. Sorushanova A., Delgada L. M., Wu Z., et al. The collagen suprafamily: From biosynthesis to advanced biomaterial development[J]. Advanced Materials, 2019, 31 (1): 1801651.
4. Xie J., Ping H., Tan T., et al. Bioprocess-inspired fabrication of materials with new structures and functions[J]. Progress in Materials Science, 2019, 105: 100571.
5. Chen J., Ahn T., Colón-Bernal I. D., et al. The relationship of collagen structural and compositional heterogeneity to tissue mechanical properties: A chemical perspective[J]. ACS Nano, 2017, 11 (11): 10665-10671.
6. Parenteau N. Skin: The first tissue-engineered products[J]. Scientific American, 1999, 280 (4): 83-84.
7. Pissarenko A., Meyers M. A. The materials science of skin: Analysis, characterization, and modeling[J]. Progress in Materials Science, 2020, 110: 52.
8. Mouw J. K., Ou G. Q., Weaver V. M. Extracellular matrix assembly: A multiscale deconstruction[J]. Nature Reviews Molecular Cell Biology, 2014, 15 (12): 771-785.
9. Campbell J. J., Husmann A., Hume R. D., et al. Development of three-dimensional collagen scaffolds with controlled architecture for cell migration studies using breast cancer cell lines [J]. Biomaterials, 2017, 114: 34-43.
10. Bendtsen S. T., Wei M. Synthesis and characterization of a novel injectable alginate-collagen-hydroxyapatite hydrogel for bone tissue regeneration[J]. Journal of Materials Chemistry B, 2015, 3 (15): 3081-3090.
11. Köster S., Evans H. M., Wong J. Y., et al. An in situ study of collagen self-assembly processes[J]. Biomacromolecules, 2008, 9 (1): 199-207.
12. Parenteaubareil R., Gauvin R., François B. Collagen-based biomaterials for tissue engineering applications[J]. Materials, 2010, 3(3): 1863-1887.
13. Liu X., Zheng M., Wang X., et al. Biofabrication and characterization of collagens with different hierarchical architectures[J]. ACS Biomaterials Science & Engineering, 2020, 6 (1): 739-748.
14. Chattopadhyay S., Raines R. T. Review collagen-based biomaterials for wound healing[J].

Biopolymers, 2014, 101 (8): 821-833.
15. Cen L., Liu W., Cui L., et al. Collagen tissue engineering: Development of novel biomaterials and applications[J]. Pediatric Research, 2008, 63: 492-496.
16. Wahyudi H., Reynolds A. A., Li Y., et al. Targeting collagen for diagnostic imaging and therapeutic delivery[J]. Journal of Controlled Release, 2016, 240: 323-331.
17. Cen L., Liu W., Cui L., et al. Collagen tissue engineering: Development of novel biomaterials and applications[J]. Pediatric Research, 2008, 63 (5): 492-496.
18. Ricardblum S. The collagen family[J]. Cold Spring Harbor Perspectives in Biology, 2010, 3 (1): a004978.
19. Kielty C. M., Grant M. E. The collagen family: Structure, assembly, and organization in the extracellular matrix, in Connective Tissue and Its Heritable Disorders: Molecular, Genetic, and Medical Aspects[M]. Wiley-Liss, Inc., 2003.
20. Ricard-Blum S., Ruggiero F. The collagen superfamily: From the extracellular matrix to the cell membrane[J]. Pathologie Biologie, 2005, 53 (7): 430-442.
21. Gordon M. K., Hahn R. A. Collagens[J]. Cell & Tissue Research, 2010, 339 (1): 247-257.
22. Engel J., Bächinger H. P. Structure, stability and folding of the collagen triple helix[J]. Topics in Current Chemistry, 2005, 247: 7-33.
23. Hulmes D. J. S. Collagen diversity, synthesis and assembly[M]. Springer, 2008.
24. Ottani V., Martini D., Franchi M., et al. Hierarchical structures in fibrillar collagens[J]. Micron, 2002, 33 (7): 587-596.
25. Yang W., Meyers M. A., Ritchie R. O. Structural architectures with toughening mechanisms in Nature: A review of the materials science of Type-I collagenous materials[J]. Progress in Materials Science, 2019, 103: 425-483.
26. Pradhan S. M., Katti K. S., Katti D. R. Structural hierarchy controls deformation behavior of collagen[J]. Biomacromolecules, 2012, 13 (8): 2562-2569.
27. Bella J. A new method for describing the helical conformation of collagen: Dependence of the triple helical twist on amino acid sequence[J]. Journal of Structural Biology, 2010, 170 (2): 377-391.
28. Harris T., Chenoweth D. M. Sterics and stereoelectronics in aza-glycine: Impact of aza-glycine preorganization in triple helical collagen[J]. Journal of the American Chemical Society, 2019, 141 (45): 18021-18029.
29. Hall D. A., Reed R. Hydroxyproline and thermal stability of collagen[J]. Nature, 1957, 180 (4579): 243.
30. Egli J., Erdmann R. S., Schmidt P. J., et al. Effect of N- and C-terminal functional groups on the stability of collagen triple helices[J]. Chemical Communications, 2017, 53(80): 11036-11039.
31. Xu S., Gu M., Wu K., et al. Unraveling the role of hydroxyproline in maintaining the thermal stability of the collagen triple helix structure using simulation[J]. The Journal of Physical

Chemistry B, 2019, 123 (36): 7754-7763.

32. Doig A. J. Statistical thermodynamics of the collagen triple-helix/coil transition. Free energies for amino acid substitutions within the triple-helix [J]. The Journal of Physical Chemistry B, 2008, 112 (47): 15029-15033.

33. Buttar D., Docherty R., Swart R. M. The application of computational chemistry to the study of the chemistry of collagen[J]. Journal of the American Leather Chemists Association, 1997, 92 (8): 185-198.

34. Persikov A. V., Ramshaw J. A. M., Kirkpatrick A., et al. Amino acid propensities for the collagen triple-helix[J]. Biochemistry, 2000, 39 (48): 14960-14967.

35. Berisio R., Granata V., Vitagliano L., et al. Imino acids and collagen triple helix stability: Characterization of collagen-like polypeptides containing Hyp-Hyp-Gly sequence repeats[J]. Journal of the American Chemical Society, 2004, 126 (37): 11402-11403.

36. Cutini M., Bocus M., Ugliengo P. Decoding collagen triple helix stability by means of hybrid DFT simulations[J]. The Journal of Physical Chemistry B, 2019, 123 (34): 7354-7364.

37. Kirkness M. W. H., Lehmann K., Forde N. R. Mechanics and structural stability of the collagen triple helix[J]. Current Opinion in Chemical Biology, 2019, 53: 98-105.

38. Mu C., Li D., Lin W., et al. Temperature induced denaturation of collagen in acidic solution [J]. Biopolymers, 2007, 86 (4): 282-287.

39. Strasser S., Zink A., Janko M., et al. Structural investigations on native collagen type I fibrils using AFM[J]. Biochemical and Biophysical Research Communications, 2007, 354 (1): 27-32.

40. Parry D. A., Craig A. S. Quantitative electron microscope observations of the collagen fibrils in rat-tail tendon[J]. Biopolymers, 2010, 16 (5): 1015-1031.

41. Xu M., Liu J., Sun J., et al. Optical microscopy and electron microscopy for the morphological evaluation of tendons: A mini review[J]. Orthopaedic Surgery, 2020, 12: 366-371.

42. Buehler M. J. Nature designs tough collagen: Explaining the nanostructure of collagen fibrils [J]. Proceedings of the National Academy of Sciences, 2006, 103 (33): 12285-12290.

43. Shoulders M. D., Raines R. T. Collagen structure and stability [J]. Annual Review of Biochemistry, 2009, 78 (1): 929-958.

44. Smith J. W. Molecular pattern in native collagen[J]. Nature, 1968, 219 (5150): 157-158.

45. Brown E. M. Development and utilization of a bovine type I collagen microfibril model[J]. International Journal of Biological Macromolecules, 2013, 53: 20-25.

46. Brown E. M., King G. Use of computer-generated models in studies of modified collagen[J]. Journal of the American Leather Chemists Association, 1996, 91 (6): 161-170.

47. Gautieri A., Vesentini S., Redaelli A., et al. Hierarchical structure and nanomechanics of collagen microfibrils from the atomistic scale up[J]. Nano Letters, 2011, 11 (2): 757-766.

48. Wess T. J. Collagen fibril form and function[J]. Advances in Protein Chemistry, 2005, 70: 341-374.

49. Bozec L., van der Heijden G., Horton M. Collagen fibrils: Nanoscale ropes[J]. Biophysical Journal, 2007, 92 (1): 70-75.
50. Erickson B., Fang M., Wallace J. M., et al. Nanoscale structure of type I collagen fibrils: Quantitative measurement of D-spacing[J]. Biotechnology Journal, 2013, 8 (1): 117-126.
51. Varma S., Orgel J. R. O., Schieber J. Nanomechanics of type I collagen[J]. Biophysical Journal, 2016, 111 (1): 50-56.
52. Fang M., Goldstein E. L., Turner A. S., et al. Type I collagen D-spacing in fibril bundles of dermis, tendon, and bone: Bridging between nano- and micro- level tissue hierarchy[J]. ACS Nano, 2012, 6 (11): 9503-9514.
53. Gayatri R., Sharma A. K., Rajaram R., et al. Chromium(III)-induced structural changes and self-assembly of collagen[J]. Biochemical and biophysical research communications, 2001, 283 (1): 229-235.
54. He L., Mu C., Shi J., et al. Modification of collagen with a natural cross-linker, procyanidin[J]. International Journal of Biological Macromolecules, 2011, 48 (2): 354-359.
55. Fratzl P., Misof K., Zizak I., et al. Fibrillar structure and mechanical properties of collagen[J]. Journal of Structural Biology, 1998, 122 (1-2): 119-122.
56. Sizeland K. H., Basil-Jones M. M., Edmonds R. L., et al. Collagen orientation and leather strength for selected mammals[J]. Journal of Agricultural and Food Chemistry, 2013, 61 (4): 887-892.
57. Meyer E. Internal water molecules and H-bonding in biological macromolecules: A review of structural features with functional implications[J]. Protein Science, 1992, 1 (12): 1543-1562.
58. Levy Y., Onuchic J. N. Water and proteins: A love-hate relationship[J]. Proceedings of the National Academy of Sciences of the United States of America, 2004, 101 (10): 3325-3326.
59. Chaplin M. Do we underestimate the importance of water in cell biology?[J]. Nature Reviews Molecular Cell Biology, 2006, 7 (11): 861-866.
60. De Simone A., Vitagliano L., Berisio R. Role of hydration in collagen triple helix stabilization[J]. Biochemical and Biophysical Research Communications, 2008, 372 (1): 121-125.
61. Spitzner E. C., Röper S., Zerson M., et al. Nanoscale swelling heterogeneities in Type I collagen fibrils[J]. ACS Nano, 2015, 9 (6): 5683-5694.
62. Bienkiewicz K. J. Leather-water a system?[J]. Journal of the American Leather Chemists Association, 1990, 85: 305-325.
63. Covington A. D. Modern tanning chemistry[J]. Chemical Society Reviews, 1997, 26(2): 111-126.
64. Covington A. D. Advances in tanning theory[J]. in IULTCS. Cape Town, 2001.
65. Ramasami T. Approach towards a unified theory for tanning: Wilson's dream[J]. Journal of the American Leather Chemists Association, 2001, 96 (8): 290-304.
66. Reich G. The structural changes of collagen during the leather making process[J]. Journal of the Society of Leather Technologists and Chemists, 1999, 83 (19): 63-79.

67. Deb Choudhury S., Haverkamp R. G., DasGupta S., et al. Effect of oxazolidine E on collagen fibril formation and stabilization of the collagen matrix[J]. Journal of Agricultural and Food Chemistry, 2007, 55 (17): 6813-6822.

68. Wu B., Mu C., Zhang G., et al. Effects of Cr^{3+} on the structure of collagen fiber[J]. Langmuir, 2009, 25 (19): 11905-11910.

69. Sizeland K. H., Edmonds R. L., Basil-Jones M. M., et al. Changes to collagen structure during leather processing[J]. Journal of Agricultural and Food Chemistry, 2015, 63 (9): 2499-2505.

70. Zhang Y., Ingham B., Cheong S., et al. Real-time synchrotron small-angle X-ray scattering studies of collagen structure during leather processing[J]. Industrial & Engineering Chemistry Research, 2018, 57 (1): 63-69.

71. Shi J., Wang C., Ngai T., et al. Diffusion and binding of Laponite clay nanoparticles into collagen fibers for the formation of leather matrix[J]. Langmuir, 2018, 34 (25): 7379-7385.

72. Cucos A., Budrugeac P. The impact of natural ageing on the hydrothermal stability of new and artificially aged parchment and leather samples[J]. Thermochimica Acta, 2018, 669: 40-44.

73. Sebestyen Z., Czegeny Z., Badea E., et al. Thermal characterization of new, artificially aged and historical leather and parchment[J]. Journal of Analytical and Applied Pyrolysis, 2015, 115: 419-427.

74. Cucos A., Budrugeac P., Miu L. DMA and DSC studies of accelerated aged parchment and vegetable-tanned leather samples[J]. Thermochimica Acta, 2014, 583: 86-93.

75. Cicero C., Mercuri F., Paoloni S., et al. Integrated adiabatic scanning calorimetry, light transmission and imaging analysis of collagen deterioration in parchment[J]. Thermochimica Acta, 2019, 676: 263-270.

76. Gayatri R., Rajaram R., Ramasami T. Inhibition of collagenase by Cr(Ⅲ): its relevance to stabilization of collagen[J]. Biochimica Et Biophysica Acta, 2000, 1524 (2): 228-237.

77. Boki K., Kawasaki N., Minami K., et al. Structural analysis of collagen fibers by nitrogen adsorption method[J]. Journal of Colloid and Interface Science, 1993, 157 (1): 55-59.

78. Fathima N. N., Kumar M. P., Rao J. R., et al. A DSC investigation on the changes in pore structure of skin during leather processing[J]. Thermochimica Acta, 2010, 501 (1-2): 98-102.

79. Fathima N. N., Dhathathreyan A., Ramasami T. Influence of crosslinking agents on the pore structure of skin[J]. Colloids and Surfaces B-Biointerfaces, 2007, 57 (1): 118-123.

80. Gil R. R., Ruiz B., Lozano M. S., et al. The role of crosslinking treatment on the pore structure and water transmission of biocollagenic materials[J]. Journal of Applied Polymer Science, 2013, 130 (3): 1812-1822.

81. He X., Wang Y., Zhou J., et al. Suitability of pore measurement methods for characterizing the hierarchical pore structure of leather[J]. Journal of the American Leather Chemists Association, 2019, 114 (2): 41-47.

82. Fathima N. N., Dhathathreyan A., Ramasami T. Mercury intrusion porosimetry, nitrogen adsorption,

and scanning electron microscopy analysis of pores in skin[J]. Biomacromolecules, 2002, 3 (5): 899-904.

83. Sivakumar V., Jena A., Gupta K., et al. Analysis of pore-size and related parameters for leather matrix through capillary flow porosimetry technique [J]. Journal of the Society of Leather Technologists and Chemists, 2015, 99 (1): 16-22.

84. Bailey A. J., Light N. D., Atkins E. D. T. Chemical cross-linking restrictions on models for the molecular organization of the collagen fibre[J]. Nature, 1980, 288 (5789): 408-410.

85. Chen J. M., Feairheller S. H., Brown E. M. Three-dimensional energy-minimized models for calfskin type I collagen triple helix and microfibril: II. the "Smith" microfibril[J]. Journal of the American Leather Chemists Association, 1991, 86 (2): 487-497.

86. Brown E. M., Chen J. M., Feairheller S. H. Predicted interactions of ionizable side chains in a fragment of the three-dimensional energy-minimized model for calf skin Type I collagen microfibril[J]. Journal of the American Leather Chemists Association, 1993, 88 (1): 2-11.

87. Chen J. M., Sheldon A., Pincus M. R. Three-dimensional energy-minimized model of human type II "Smith" collagen microfibril[J]. Journal of Biomolecular Structure & Dynamics, 1995, 12 (6): 1129-1159.

88. Covington A. D. New tannages for the new millennium[J]. Journal of the American Leather Chemists Association, 1998, 93 (6): 168-182.

89. Covington A. D. Chrome tanning: Exploding the perceived myths, preconceptions and received wisdom[J]. Journal of the American Leather Chemists Association, 2001, 96 (12): 467-480.

90. Ramasami T. Approach towards a unified theory for tanning: Wilson's dream[J]. Journal of the American Leather Chemists Association, 2001, 96: 290-304.

91. Gustavson K. H. The chemistry of tanning processes[J]. Science, 1956, 124 (3210): 36-37.

92. Gustavson K. H. The Chemistry and Reactivity of Collagen [M]. New York: Academic Press, 1956.

93. Kühn K. The classical collagens: Types I, II, and III, in Structure and Function of Collagen Types[M]. R. Mayne and R. E. Burgeson, Editors. Academic Press, 1987: 1-42.

94. Bailey A. J., Paul R. G. Collagen: A not so simple protein[J]. Journal of the Society of Leather Technologies and Chemists, 1998, 82: 104-110.

95. Covington A. D., Shi B. High stability organic tanning using plant polyphenols. Part 1: the interactions between vegetable tannins and aldehydic crosslinkers[J]. Journal of the Society of Leather Technologists and Chemists, 1998, 82 (2): 64-71.

96. Brown E. M., Dudley R. L., Elsetinow A. R. A conformational study of collagen as affected by tanning procedures[J]. Journal of the American Leather Chemists Association, 1997, 92 (9): 225-232.

97. Reich T., Rossgerg A., Hennig C., et al. Characterization of chromium complexes in chrome tannins, leather, and gelatin using extended X-ray absorption fine structure (EXAFS)

spectroscopy[J]. Journal of the American Leather Chemists Association, 2001, 96(4): 133-147.

98. Stachel I., Schwarzenbolz U., Henle T., et al. Cross-linking of type I collagen with microbial transglutaminase: identification of cross-linking sites[J]. Biomacromolecules, 2010, 11 (3): 698-705.

99. Wu X., Liu Y., Liu A., et al. Improved thermal-stability and mechanical properties of type I collagen by crosslinking with casein, keratin and soy protein isolate using transglutaminase [J]. International Journal of Biological Macromolecules, 2017, 98: 292-301.

100. Covington A. D. Theory and mechanism of tanning: Present thinking and future implications for industry[J]. Journal of the Society of Leather Technologists & Chemists, 2001, 85 (1): 24-34.

101. Bienkiewicz K. J. Physical chemistry of leather making[M]. Malabar, Florida: R O. Krieger Publishing Co., 1980.

102. Suresh V., Kanthimathi M., Thanikaivelan P., et al. An improved product-process for cleaner chrome tanning in leather processing[J]. Journal of Cleaner Production, 2001, 9 (6): 483-491.

103. Sundar V. J., Raghava Rao J., Muralidharan C. Cleaner chrome tanning-emerging options[J]. Journal of Cleaner Production, 2002, 10 (1): 69-74.

104. Sundar V. J., Muralidharan C., Mandal A. B. A novel chrome tanning process for minimization of total dissolved solids and chromium in effluents [J]. Journal of Cleaner Production, 2013, 59: 239-244.

105. Wolf G., Breth M., Carle J., et al. New developments in wet-white tanning technology[J]. Journal of the American Leather Chemists Association, 2001, 96 (4): 111-120.

106. Onem E., Yorgancioglu A., Karavana H. A., et al. Comparison of different tanning agents on the stabilization of collagen via differential scanning calorimetry [J]. Journal of Thermal Analysis and Calorimetry, 2017, 129 (1): 615-622.

107. Miles C. A., Bailey A. J. Thermal denaturation of collagen revisited[J]. Proceedings of the Indian Academy of Sciences-Chemical Sciences, 1999, 111 (1): 71-80.

108. Komanowsky M. Thermal stability of hide and leather at different moisture content [J]. Journal of the American Leather Chemists Association, 1991, 86 (8): 269-280.

109. Komanowsky M. Thermodynamic analysis of thermal denaturation of hide and leather[J]. Journal of the American Leather Chemists Association, 1992, 87 (2): 52-66.

110. Kronick P. L., Cooke P. Thermal stabilization of collagen fibers by calcification [J]. Connective Tissue Research, 1996, 33 (4): 275-282.

111. Fathima N. N., Dhathathreyan A., Ramasami T. A new insight into the shrinkage phenomenon of hides and skins[J]. Journal of the American Leather Chemists Association, 2001, 96 (11): 417-425.

112. Taylor M. M., Cabeza L. F., Marmer W. N., et al. Enzymatic modification of hydrolysis products from collagen using a microbial transglutaminase. I Physical properties[J]. Journal of the American Leather Chemists Association, 2001, 96 (9): 319-332.

113. Reich G. Scanning probe microscopy—A useful tool in leather research[J]. Journal of the Society of Leather Technologists and Chemists, 1998, 82 (1): 11-14.
114. Covington A. D., Lampard G. S., Menderes O., et al. Extended X-ray absorption fine structure studies of the role of chromium in leather tanning[J]. Polyhedron, 2001, 20 (5): 461-466.
115. Romer F. H., Underwood A. P., Senekal N. D., et al. Tannin fingerprinting in vegetable tanned leather by solid state NMR spectroscopy and comparison with leathers tanned by other processes[J]. Molecules, 2011, 16 (2): 1240-1252.
116. Zhang Y., Snow T., Smith A. J., et al. A guide to high-efficiency chromium(III)-collagen cross-linking: Synchrotron SAXS and DSC study[J]. International Journal of Biological Macromolecules, 2019, 126: 123-129.
117. Rodriguez Gil R., Ruiz B., Santiago Lozano M., et al. The role of crosslinking treatment on the pore structure and water transmission of biocollagenic materials[J]. Journal of Applied Polymer Science, 2013, 130 (3): 1812-1822.
118. Bosch T., Manich A. M., Carilla J., et al. Collagen thermal transitions in chrome leather-Thermogravimetry and differential scanning calorimetry[J]. Journal of the American Leather Chemists Association, 2002, 97 (11): 441-450.
119. Miles C. A. Kinetics of the helix/coil transition of the collagen-like peptide (Pro-Hyp-Gly)$_{10}$[J]. Biopolymers, 2010, 87 (1): 51-67.
120. Scheraga H. From helix-coil transitions to protein folding[J]. Biopolymers, 2010, 89 (5): 479-485.
121. Sizeland K. H., Wells H. C., Kelly S. J. R., et al. The influence of water, lanolin, urea, proline, paraffin and fatliquor on collagen D-spacing in leather[J]. RSC Advance, 2017, 7 (64): 40658-40663.
122. Choudhury S. D., DasGupta S., Norris G. E. Unravelling the mechanism of the interactions of oxazolidine A and E with collagens in ovine skin[J]. International Journal of Biological Macromolecules, 2007, 40 (4): 351-361.
123. Khanbabaee K., van Ree T. Tannins: Classification and definition[J]. Natural Product Reports, 2001, 18 (6): 641-649.
124. Sieniawska E., Baj T. Chapter 10-Tannins, in Pharmacognosy[M]. S. Badal and R. Delgoda, Editors. Academic Press: Boston, 2017: 199-232.
125. Bentley W. E., Payne G. F. Nature's other self-assemblers[J]. Science, 2013, 341 (6142): 136-137.
126. Deaville E. R., Green R. J., Mueller-Harvey I., et al. Hydrolyzable tannin structures influence relative globular and random coil protein binding strengths[J]. Journal of Agricultural and Food Chemistry, 2007, 55 (11): 4554-4561.
127. Shin M., Park E., Lee H. Plant-inspired pyrogallol-containing functional materials[J]. Advanced Functional Materials, 2019, 29: 1903022.

128. Marais J. P. J., Mueller-Harvey I., Brandt E. V., et al. Polyphenols, condensed tannins, and other natural products in onobrychis viciifolia (Sainfoin) [J]. Journal of Agricultural and Food Chemistry, 2000, 48 (8): 3440-3447.
129. Duval A., Avérous L. Characterization and physicochemical properties of condensed tannins from Acacia catechu[J]. Journal of Agricultural and Food Chemistry, 2016, 64 (8): 1751-1760.
130. Chai W. M., Huang Q., Lin M., et al. Condensed tannins from Longan Bark as inhibitor of tyrosinase: Structure, activity, and mechanism[J]. Journal of Agricultural and Food Chemistry, 2018, 66 (4): 908-917.
131. Madhan B., Subramanian V., Rao J. R., et al. Stabilization of collagen using plant polyphenol: Role of catechin[J]. International Journal of Biological Macromolecules, 2005, 37 (1): 47-53.
132. Shirmohammadli Y., Efhamisisi D., Pizzi A. Tannins as a sustainable raw material for green chemistry: A review[J]. Industrial Crops and Products, 2018, 126: 316-332.
133. Nashy E. H. A., Osman O., Mahmoud A. A., et al. Molecular spectroscopic study for suggested mechanism of chrome tanned leather[J]. Spectrochimica Acta Part A: Molecular and Biomolecular Spectroscopy, 2012, 88: 171-176.
134. Covington A., Lampard G. S., Hancock R. A., et al. Studies on the origin of hydrothermal stability: A new theory of tanning[J]. Journal of the American Leather Chemists Association, 1998, 93: 107-120.
135. Heth C. L. The skin they were in: Leather and tanning in antiquity, in Chemical Technology in Antiquity[M]. American Chemical Society, 2015: 181-196.
136. Zengin A. C. A., Crudu M., Maier S. S., et al. Eco-leather: Chromium-free leather production using titanium, oligomeric melamine-formaldehyde resin, and resorcinol tanning agents and the properties of the resulting leathers[J]. Ekoloji, 2012, 21 (82): 17-25.
137. Crudu M., Deselnicu V., Deselnicu D. C., et al. Valorization of titanium metal wastes as tanning agent used in leather industry[J]. Waste Management, 2014, 34 (10): 1806-1814.
138. Peng B., Shi B., Ding K., et al. Novel titanium (IV) tanning for leathers with superior hydrothermal stability[J]. Journal of the American Leather Chemists Association, 2007, 10 (10): 297-305.
139. Christian S., Pomelli C. C., Federica B., et al. An insight into the molecular mechanism of the masking process in titanium tanning[J]. Clean Technologies and Environmental Policy, 2017, 19: 259-267.
140. Sundarrajan A., Madhan B., Rao R., et al. Studies on tanning with zirconium oxychloride: Part I: Standardization of tanning process[J]. Journal of the American Leather Chemists Association, 2003, 98 (3): 101-106.
141. Cao S., Zeng Y., Cheng B., et al. Effect of pH on Al/Zr-binding sites between collagen fibers in tanning process[J]. Journal of the American Leather Chemists Association, 2016, 111 (7): 242-249.

142. Suparno O., Kartika I. A., Muslich. Chamois leather tanning using rubber seed oil[J]. Journal of the Society of Leather Technologists and Chemists, 2009, 93 (4): 158-161.
143. Sandhya K. Y., Vedaraman N., Sundar V. J., et al. Suitability of different oils for chamois leather manufacture[J]. Journal of the American Leather Chemists Association, 2015, 110 (7): 221-226.
144. Sundar V. J., Muralidharan C., Mandal A. B. Eco-benign stabilization of skin protein-Role of Jatropha curcas oil as a co-tanning agent[J]. Industrial Crops and Products, 2013, 47: 227-231.
145. Wainaina P. N., Tanui P., Ongarora B. Extraction of oil from tannery fleshings for chamois leather tanning[J]. Journal of the Society of Leather Technologists and Chemists, 2019, 103 (3): 159-162.
146. Tian Z., Liu W., Li G. The microstructure and stability of collagen hydrogel cross-linked by glutaraldehyde[J]. Polymer Degradation and Stability, 2016, 130: 264-270.
147. Mesquida P., Kohl D., Andriotis O. G., et al. Evaluation of surface charge shift of collagen fibrils exposed to glutaraldehyde[J]. Scientific Reports, 2018 (1): 10126.
148. Deb Choudhury S., Haverkamp R. G., DasGupta S., et al. Effect of oxazolidine on collagen fibril formation and stabilization of the collagen matrix[J]. Journal of Agricultural and Food Chemistry, 2007, 55 (17): 6813-6822.
149. Okoro C. C., Samuel O., Lin J. The effects of tetrakis-hydroxymethyl phosphonium sulfate (THPS), nitrite and sodium chloride on methanogenesis and corrosion rates by methanogen populations of corroded pipelines[J]. Corrosion Science, 2016, 112: 507-516.
150. Xu D., Li Y., Gu T. A synergistic D-tyrosine and tetrakis hydroxymethyl phosphonium sulfate biocide combination for the mitigation of an SRB biofilm[J]. World Journal of Microbiology & Biotechnology, 2012, 28 (10): 3067-3074.
151. Covington A. D. Other tannages. In Tanning Chemistry: The Science of Leather[M]. Royal Society of Chemistry: 2009.
152. Chung C., Lampe K. J., Heilshorn S. C. Tetrakis (hydroxymethyl) phosphonium chloride as a covalent cross-linking agent for cell encapsulation within protein-based hydrogels [J]. Biomacromolecules, 2012, 13 (12): 3912-3916.
153. Li Y., Shao S., Shi K., et al. Release of free formaldehyde in THP salt tannages[J]. Journal of the Society of Leather Technologists and Chemists, 2008, 92 (4): 167-169.
154. Oehmen A., Lemos P. C., Carvalho G., et al. Advances in enhanced biological phosphorus removal: From micro to macro scale[J]. Water Research, 2007, 41 (11): 2271-2300.
155. Aravindhan, K., Madhan, B., and Rama, K. Stndies on tara-phosphonium Combination tannage: Approach towards a metal free eco-benign tanning system [J]. Journal of the American Leather chemists Assoclution, 2015, 110 (3): 80-87.
156. Ren L., Wang X., Qiang T., et al. Phosphonium-aluminum combination tanning for goat garment leather[J]. Journal of the American Leather Chemists Association, 2009, 104 (6):

218-226.

157. Shi J., Wang C., Hu L., et al. Novel wet-white tanning approach based on Laponite clay nanoparticles for reduced formaldehyde release and improved physical performances[J]. ACS Sustainable Chemistry & Engineering, 2019, 7 (1): 1195-1201.

158. Jaisankar S. N., Gupta S., Lakshminarayana Y., et al. Water-based anionic sulfonated melamine formaldehyde condensate oligomer as retanning agent for leather processing[J]. Journal of the American Leather Chemists Association, 2010, 105 (9): 289-296.

159. Chen M., Wang Y., Fan H., et al. Particle size evolution of melamain-formaldehyde tanning agent on tanning effect[J]. Journal of the American Leather Chemists Association, 2018, 113 (5): 151-162.

160. Bai X., Chang J., Chen Y., et al. A novel chromium-free tanning process based on *in-situ* melamine-formaldehyde oligomer condensate[J]. Journal of the American Leather Chemists Association, 2013, 108 (11): 404-410.

Appendix I

(According to "ISO/FDIS 15115 Leather-Vocabulary", "EN 15987 Leather-Terminology-Key definitions for the leather trade" and "BS 2780: 1983 + A1: 2013 Glossary of leather terms")

Process

basification

Mild alkali treatment to ensure completion of tanning.

bating

Removing unwanted interfibrillary proteins by treating the hides and skins with bates, to obtain soft and flexible leather with a smooth grain surface.

curing

Temporary preservation of raw hides and skins.

degreasing

Removing natural fat from the hide or skin by emulsification in an aqueous media and/or using a solvent media.

deliming

Removing the alkalinity of limed pelt.

fatliquoring

Application of fat liquors for lubricating and softening leathers.

finishing

Chemical and/or mechanical operations carried out on crust leathers to impart the desired properties for the intended final use of the leather.

liming

Treating raw hides and skins with lime liquor with a view to plumping and/or unhairing.

neutralization

Raising the pH of a mineral-tanned leather towards neutral from an acidic zone by treatment with a solution of salt of a weak alkali or buffer mixture.

pickling

Treating the delimed pelts with acid and salt to lower the pH.

sammying

Reducing the moisture content of pelts or leathers by squeezing between the rollers in a machine.

shaving

Mechanical operation carried out to make the thickness of the leather uniform.

soaking

Treatment with water to clean the skin and get it back to the original condition.

tanning

Treatment of hide or skin with extracts of natural products (e. g. bark, leaves, seeds) or chemical agents (e. g. chromium, aluminium, organic compounds) to stabilize against heat, enzymatic attack and thermo-mechanical stress.

tawing

The processing of hides and skins with alum and salt.

Material

aldehyde leather

Washable leather, which in its natural state is white/cream, prepared usually from sheepskin or lambskin splits or degrains and tanned with an aldehyde.

aniline leather

Leather whose natural grain is clearly visible either without a surface coating or with a non-pigmented surface coating of thickness not exceeding 0.01 mm.

belly

Part of the hide covering the underside and the upper part of the legs of the animal.

belt leather

Leather used for waist belts as distinct from transmission belting.

binder

Film-forming material, usually polymeric, used to adhere pigment particles and additives for coating the surface of the leather.

bovine leather

Leather made from the hide or skin of a bovine animal, usually applied to the hide of an ox or cow.

buffed leather

Leather from which some of the grain has been removed by an abrasive or bladed cylinder or, less generally, by a hand tool.

chamois leather

Leather made from the flesh split of sheep, goat or lamb skin, or from sheep or lamb skin from which the grain has been removed and tanned by processes involving the oxidation of fish or marine oils.

Fundamental Collagen Chemistry in Leather Making

chrome-tanned leather

Hide or skin converted to leather either by treatment solely with chromium salts or with chromium salts together with a small amount of some other tanning agent, used merely to assist the chromium tanning process, and not in sufficient amount to alter the essential chromium tanned character of the leather.

chrome-free leather

Hide or skin converted to leather by a tanning agent free of chromium salts, where the total content of chromium in the tanned leather is less than or equal to 0.1% (mass of chromium/total dry weight of leather).

coated leather

A product where the surface coating applied to the leather substrate does not exceed one-third of the total thickness of the product, but is in excess of 0.15 mm.

combination tanned leather

Leather tanned by two or more tanning agents, e. g. chrome followed by vegetable (chrome re-tan), vegetable followed by chrome (semi-chrome) or formaldehyde followed by oil (combination).

corrected grain leather

Leather from which the grain layer has been partially removed by buffing to a depth governed by the condition of the raw material and upon which a new surface has been built by various finishes.

crust leather

Leather which is tanned, fat liquored and dried before finishing.

fatliquor

Natural or synthetic oil-based formulations that are emulsions, solutions or dispersions used to lubricate the fibers of leather.

embossed leather

Leather embossed or printed with a three-dimensional pattern either imitating or resembling the grain pattern of some animal, or unrelated to a natural grain pattern.

full grain

Leather bearing the original grain surface as exposed by removal of the epidermis and with none of the surface removed by buffing, snuffing or splitting.

grain

Outer side of the leather once the hair or wool and epidermis have been removed, characterized by pores from hair or wool, feather follicles or scales, specific to each animal species.

grain split

Outer grain layer of a hide or skin separated from the inner layer(s) by splitting horizontally in a machine.

hide

Raw skin of a mature or fully-grown animal of the larger kind.

hide powder

Powder from well-washed, dried, delimed pelt disintegrated using a grinding mill.

impregnated leather

Leather that, by means of the penetration of materials, such as grease, wax and/or impregnating resins, has been improved in regard to certain of its physical properties.

laminated leather

A composite of a layer of leather and one or more layers of another sheet or film of plastics or other material.

leather

Hide or skin with its original fibrous structure more or less intact, tanned to be imputrescible, where the hair or wool may or may not have been removed, whether or not the hide or skin has been split into layers or segmented either before or after tanning and where any surface coating or surface layer, however applied, is not thicker than 0.15 mm.

masking agent

Salts of weak acids added during mineral tanning to prevent precipitation of tanning salts.

metal-free leather

Hide or skin converted to leather by a tanning agent free of metallic salts (Cr, Al, Ti, Zr, Fe), where the total content of each tanning metal in the tanned leather is less than or equal to 0.1% (mass of each metal/total dry weight of unfinished leather).

Nappa

Soft full-grain leather, formerly made from unsplit sheepskin or lambskin or kidskin for gloving and clothing, but nowadays also made from split hide. It was originally tanned with chromium salts and dyed throughout its substance.

nubuck

Cattle-hide leather buffed on the grain side to give a very fine velvety surface, white or colored.

parchment

Dry, translucent or opaque untanned hide or skin material.

patent leather

Leather with a lustrous mirror-like surface, built up by the application of one or more coats of varnishes or lacquers pigmented or non-pigmented, whose thickness doesn't exceed one-third of the total thickness.

pelt

Hide and skin prepared for tanning by removal of the flesh and, generally, hair or wool.

semi-aniline leather

Leather applied with a finishing coat using a small quantity of pigments so that the original grain pattern is not concealed.

semi-chrome leather

Leather which has been tanned first with vegetable tannin and then retanned with chromium salts.

shrunken grain leather

Leather specially tanned so as to shrink the grain layer, with a grain surface of prominent but uneven folds and valleys.

skin

Outer covering of smaller types of animals, e. g. sheep and goats, or of the immature animals of the larger species, e. g. calves.

split leather

Layer from a hide or skin made from a flesh split or a middle split, without any grain structure, tanned to be imputrescible.

suede

Leather whose wearing surface has been finished to produce a velvet-like nap.

vegetable-tanned leather

Hide or skin converted to leather by vegetable tanning agents, where the total content of tanning metals (Cr, Al, Ti, Zr, Fe) is less than or equal to 0.3% (mass of all metals/total dry weight of leather).

wet-blue

Wet chrome tanned hide or skin which is an intermediate material of leather manufacturing.

wet-white

Hide or skin tanned with substances such as aldehyde, aluminium and syntans, conferring white color and in wet condition, which is an intermediate material of leather manufacturing.

water-resistant leather

Leather resistant to the penetration of water, usually chrome tanned or combination tanned, originally heavily greased but nowadays other water repelling agents may be used.

Equipment

beam

Convex wooden slab sloping downward from about waist height over which a hide is placed for unhairing, trimming off excess flesh and ragged edges, and scudding by hand knife.

drum

Cylindrical vessel with baffles inside and capable of rotation about its own axis, used for mechanical agitation in leather processing.

Performance

abrasion resistance

Ability of the leather to withstand surface wear from rubbing, chafing and other frictional forces.

break of leathers

Surface wrinkles formed when the leather is bent, grain inward.

burst strength

Force required to rupture leather.

cold crack resistance

Resistance of leather finish to crack and peel when subject to bending/flexing under a temperature not exceeding 5 ℃.

crocking

Transfer of colorant to the contact fabric when dry rubbed or wet rubbed.

flexural endurance

Ability of finished leather to endure the stress applied due to repeated bending.

fog resistance

Resistance to the release of semi-volatile and low-volatile substances present in the leather at high ambient temperature.

grain crack resistance

Resistance of the grain surface of the leather to rupture when subjected to mechanical stress.

rub fastness

Fastness to rubbing, with a felt pad, either dry or wet.

shrinkage temperature

Temperature at which an untanned skin or leather immersed in a water bath starts shrinking when heated uniformly.

Appendix Ⅱ

(The word-forming patterns for the vocabularies related to leather-making and leather chemicals)

Prefix

(1) di-表示"二、重"

dichromic acid	重铬酸
dimethylamine	二甲胺
dimethylamine unhairing	二甲胺法脱毛
diatomic acid (base)	二价酸(碱)
divalence	二价
dipeptide	二肽
disulfide	二硫化物
disulfide bond	双硫键
carbon dioxide	二氧化碳
dichromate	重铬酸盐

(2) tri-表示"三"

trimethylamine	三甲胺
trivalent	三价的
trivalence	三价
tricolour, tricolor	三元色
tripolymer	三聚体

(3) tetra-表示"四", 元音前用 tetr

tetrachloride carbon	四氯化碳
tetrachloroethylene	四氯乙烯
tetramolecular	四分子的
tetravalent alcohol	四元醇

(4) penta-表示"五"

pentachloride	五氯化物
pentahydrate	五水合物
pentachlorophenol	五氯苯酚
pentapeptide	五肽
pentanuclear	五核的、五环的

(5) hydro-表示"含水的、氢的、氢化的"

hydrogen	氢
hydrogen bond	氢键
hydrosulfide	硫氢化物
hydrochloric acid	盐酸
hydrolysis	水解
hydrometer	比重计
hydrophile	亲水物、亲水试剂
hydroxide	氢氧化物
chrome hydroxide	氢氧化铬
hydroxide ion	氢氧根离子
hydroxyl group	羟基
hydrolysable tannin	水解类单宁

(6) oxid-表示"氧、含氧的"

oxidable	可氧化的
oxidant/oxidizer/oxidizing agent	氧化剂
oxidation	氧化作用
oxide	氧化物
oxidation-reduction indicator (method)	氧化还原指示剂(法)

(7) reduc-表示"还原"

reduce	还原
reduced chrome tanning liquor	还原了的铬鞣液
reducer/reducing agent	还原剂
reduction	还原反应

(8) chlor- 表示"含氯的、氯"

chlorine	氯、氯气
chlorine dioxide	二氧化氯
chloride	氯化物
chloride ion	氯离子
bichloride	二氯化物
monochloride	一氯化物
hydrochloric acid	盐酸
chloroform	氯仿

(9) sulf- (sulfur-, sulph-, sulphur-) 表示"硫、硫磺、含硫的"

sulfuric acid	硫酸
sulfate	硫酸盐
sulfate ion	硫酸根
calcium sulfate	硫酸钙
sulfide	硫化物
sodium sulfide	硫化钠
sulfite	亚硫酸盐
sulfoacid, sulfonic acid	磺酸
sulfonate	磺酸盐
sulfonic acid group	磺酸基

(10) de- 表示"脱除、去除"

dehydration	脱水、干燥
deoxidize	脱氧、去氧
decomposition	(化学)分解(作用)
dehydrogenation	脱氢
delime	脱灰
depickle	脱酸
dechroming	脱铬
decolor	脱色
degrain	去粒面
degrease, defat	脱脂
de-alkalization	脱碱

(11) hypo-表示"次",在无机盐中表示酸和盐中心多价正性元素是处于最低氧化态

hypochlorous acid	次氯酸
sodium hypochlorite	次氯酸钠

(12) carbo-表示"碳、含碳的、与碳有关的"

carbonate	碳酸盐
calcium carbonate	碳酸钙
carbonate ion	碳酸根
carbonic acid	碳酸
carboxyl group	羧基
carboxylic acid	羧酸
carbonyl group	羰基
sodium carbonate	碳酸氢钠

(13) nitr-表示"氮、与氮有关的"

nitric acid	硝酸
nitrate	硝酸盐
sodium nitrate	硝酸纳
nitrate ion	硝酸根
nitride	硝化物
nitrite	亚硝酸盐
sodium nitrite	亚硝酸钠
nitrite ion	亚硝酸根

(14) solut-, solv-表示"与分解、溶解有关的"

solubility	溶解度、溶解性、可溶性
soluble	可溶的
insoluble	不可溶的
solute	溶质、溶解物
solvent	溶剂
solution	溶液

(15) poly-表示"多、聚合、多数"

polyacid	多元酸(有机酸)、多酸的
polyatomic	多原子的
polymer	聚合物

polymerize	聚合
polymerization	聚合(反应)
polypeptide	多肽
polyacrylic acid	聚丙烯酸
polyurethane	聚氨酯

(16) pre-表示"预处理"

presoaking	预浸水
pre-fleshing	预去肉
pre-tanning	预鞣
pretanning agent	预鞣剂
pretannage	预鞣法
preheat	预热
pretreatment	预处理
preplate	预熨烫
prebottom	预底涂层
pregrounds	预底涂

(17) centi-表示"百分之一、厘"

centimeter (re)	厘米
centigrade	百分度的
centigrade thermometer	摄氏温度计

(18) milli-表示"千分之一、毫"

millimeter	毫米
milliliter	毫升
milligram	毫克

(19) kilo-表示"千"

kilogram	千克
kilometer	千米
kiloliter	千升

Suffix

(1) -ate 表示"(正)盐、某酸盐"

chromate	铬酸盐
dichromate	重铬酸盐
sulfate (sulphate)	硫酸盐
chromium sulfate	硫酸铬
nitrate	硝酸盐
sodium nitrate	硝酸钠
carbonate	碳酸盐
calcium carbonate	碳酸钙
borate	硼酸盐
acetate	醋酸盐
sodium acetate	醋酸纳
sodium formate	甲酸钠

(2) -ate 表示"某某化作用、使化合"

oxygenate	(使)氧化
chlorinate	(使)氯化
nitrate	硝化
carbonate	碳化

(3) -ite 表示"亚盐"

chromite	亚铬酸盐
sulfite	亚硫酸盐
sodium sulfite	亚硫酸纳
sodium bisulfite	亚硫酸氢钠
nitrite	亚硝酸盐

(4) -ic 表示"某某酸"

acetic acid	醋酸
formic acid	甲酸
nitric acid	硝酸
oxalic acid	草酸
carbonic acid	碳酸

sulfuric acid	硫酸
hydrochloric acid	盐酸
chromic acid	铬酸
dichromic acid	重铬酸
aspartic acid	天冬氨酸
glutamic acid	谷氨酸

(5) -ide 表示"某某化物"

chloride	氯化物
sodium chloride	氯化纳
iodide	碘化物
oxide	氧化物
calcium oxide	氧化钙(石灰)
calcium hydroxide	氢氧化钙
chromium hydroxide	氢氧化铬
sulfide	硫化物
sodium sulfide	硫化钠
hydrogen sulfide	硫化氢

(6) -ize 表示"某某化(作用)"

sulfurize	硫化
oxidize	氧化
hydrogenize	氢化
stabilize	稳定化
ionize	离子化
neutralize	中和

(7) -ase 表示"某某酶"

peptidase	肽酶
keratinase	角蛋白酶
collagenase	胶原蛋白酶
hydrolase	水解酶

Continued

amylase	淀粉酶
oxidase	氧化酶
lipase	脂肪酶

(8) -ium 表示金属元素名称

chromium	铬
aluminium	铝
zirconium	锆
potassium	钾
magnesium	镁
titanium	钛

(9) -anol 表示"某某醇"

methanol	甲醇
ethanol	乙醇
propanol	丙醇

(10) -yl 表示"某某基团"

hydroxyl	羟基
carboxyl	羧基
carbonyl	羰基
propyl	丙基
phenyl	苯基

(11) -ed 表示经过某某工序处理过的皮或革

soaked skin or hide	浸过水的皮
limed hide	灰皮
unhaired skin	已脱毛的生皮
delimed skin	脱灰皮
bated skin	软化皮
pickled hide or skin	浸酸皮
chrome tanned leather	铬鞣革
colored/dyed leather	已染色的皮革
corrected grain leather	修面革
finished leather	成革

Continued

finished goods/product	成品
finished shoe	成品鞋
buffed grain	磨过的粒面
buffed leather	磨面革、绒面革
buffered solution	缓冲溶液
shaved leather	削匀革
shaved weight	削匀革重
sammed weight	挤水革重
wringed leather weight	挤水革重

(12) -bility 表示某种特性

solubility	溶解度、溶解性
durability	耐用性
permeability	渗透性
ability	能力
stability	稳定性
water vapour permeability	透水汽性
hydrothermal stability	湿热稳定性
compatibility	相容性、兼容性
flexibility	柔韧性、灵活性
polishability	(可)抛光性
repolishability	涂层的再抛光性
breathability	透气性
availability	实用性、可用性
dyeability	可染色性
combinability	结合性
pliability	柔韧性
embossability	压花性能
crosslinkability	交联性能

(13) -ness 表示皮革的某种性质

softness	柔软性
light fastness	耐光性
tightness	紧实性

Continued

grain tightness	粒面的紧实性
fineness of grain	粒面的细致性
firmness	紧实性
wet/dry rub fastness	耐干/湿擦性
evenness of dye	染色的均匀性
fullness	紧实性
dry cleaning and washing fastness	耐干湿洗坚牢度
hardness	硬度

(14) -er 表示"某物、某设备、某设备操作工"

splitter	片皮机(片皮操作工)
shaver	削匀机(削匀操作工)
flesher	去肉机(去肉操作工)
roller-coater	辊涂
reverse roller-coating	逆向辊涂
supplier	厂商、供应商
neutralizer	中和剂
plasticizer	增塑剂
stabilizer	稳定剂
foam-stabilizer	发泡稳定剂
emulsifier	乳化剂
thickener	增稠剂
filter paper	滤纸
promoter	促进剂、助催化剂
binder	交联剂、成膜剂
crosslinker	交联剂
foamer	发泡剂
filler	填充剂
protein filler	蛋白填料
producer	生产商、厂商
manufacturer	制造商
employer	雇员

(15) -tion/ation 表示某种动作或行为的名词形式

penetration	渗透
agitation	搅拌
precipitation	沉淀
dilution	稀释
titration	滴定
concentration	浓缩、浓度
neutralization	中和
combination tannage	结合鞣
deposition	沉积
modification	改性
migration	(颜色的)迁移
filtration	过滤
evaporation	蒸发
solubilization	溶解性
fixation	固定、结合
fixation of dye	固色
lubrication	润滑
distribution of dye	染料的分散
saturation	饱和(度)
evaluation of leather	对皮革的评价
ionization	离子化

(16) -y 表示皮革的某种感觉

waxy feel	蜡感
slippery feel	滑爽感
oily feel	油腻感
oily to dry feel	油腻至干燥的手感
strong oily greasy feel	强油腻感
silky gloss feel	丝光感
silky feel	丝绸感
round and full feel	圆润丰满的手感

(17) -sion 表示某种材料的形态或行为

suspension	悬浮液
emulsion	乳液
dispersion	分散
diffusion	扩散
adhesion	黏着力(性)、附着性

(18) -en 表示对皮革的某种作用

soften the leather	使皮革柔软
harden the leather	使皮革变硬
darken by hot plating	皮革涂层高温熨烫,颜色变深
tighten the grain	使粒面紧实
thicken	增厚、增稠
sharpen	加强、增强
lighten	(颜色)变浅

(19) -ity 表示某种特性、性质

chemical reactivity	化学反应活性
viscosity of the finish	稠度、黏性(涂饰剂的稠度)
low viscosity	低黏度
chemical activity	化学活性
relative humidity	相对湿度
thermplasticity of thin film	(涂饰剂)薄膜的热塑性
elasticity	弹性

(20) -able;-ble 表示"能……或可……"

crosslinkable	可交联的
miscible	易混合的
dilutable	可稀释的
polishable	可抛光的
glazeable	可打光的
millable	可摔软的
flexible	柔软的、柔韧的
be stable to hard water	对硬水稳定

(21) -ion 表示与离子相关的词

ionic	离子的
ionize	离子化
ionization	离子化(名词)
cationic	阳离子的
anionic	阴离子的
cation	阳离子
anion	阴离子

(22) free (组词成分), 表示没有某种化学品的皮革处理过程

chrome free tanning	无铬鞣
solvent free	不含溶剂的
ammonium-free delime and bating	无氨脱灰和软化
mineral free	不含任何金属

(23) power (组词成分), 表示某种处理皮革的能力

filling power	涂饰剂或复鞣剂的填充能力
covering power	(涂层的)遮盖力
emulsifying power	乳化力

(24) agent (组词成分), 表示某种皮革化学品(皮化材料)

waxy feel agent	蜡感助剂
anti-sticking agent	防黏剂
tanning agent	鞣剂
chrome tanning agent	铬鞣剂
retanning agent	复鞣剂
sharpening agent	加强剂
neutralizing agent	中和剂
filling agent	填充剂
feeling agent	手感剂
leveling (levelling) agent	匀染剂, 流平剂
masking agent	蒙面剂, 隐匿剂
anti-foaming agent	防泡剂
retarding agent	阻燃剂
dyestuff fixing agent	固色剂
auxiliary agent	助剂

emulsifying agent	乳化剂
degreasing agent	脱脂剂
dispersing agent	分散剂
thickening agent	增稠剂
pretanning agent	预鞣剂
reducing agent	还原剂
oxidizing agent	氧化剂
foaming agent	发泡剂
film-forming agent	成膜剂

(25) -ant 表示某种材料、试剂

oxidant	氧化剂
surfactant	表面活性剂
reductant	还原剂
redardant	阻燃剂
colorant	颜料膏
titrant	滴定剂
penetrant	渗透剂
precipitant	沉淀剂

Appendix III

Representative Research Articles

1. **Effects of Cr^{3+} on the structure of collagen fiber**

   Reprinted with permission from *Langmuir*, 2009, 25 (19): 11905-11910. Copyright (2009) American Chemical Society.

2. **Trivalent chromium and aluminum affect the thermostability and conformation of collagen very differently**

   Reprinted with permission from *Journal of Inorganic Biochemistry*, 2012, 117: 124-130. Copyright (2012) Elsevier.

3. **A comprehensive evaluation of physical and environmental performances for wet-white leather manufacture**

   Reprinted with permission from *Journal of Cleaner Production*, 2016, 139: 1512-1519. Copyright (2016) Elsevier.

4. **A novel approach for wet-white leather manufacture based on tannic acid-Laponite nanoclay combination tannage**

   Reprinted with permission from *Journal of the Society of Leather Technologists and Chemists*, 2016, 100(1): 25-30.

5. **A novel approach for lightfast wet-white leather manufacture based on sulfonesyntan-aluminum tanning agent combination tannage**

   Reprinted with permission from *Journal of the American Leather Chemists Association*, 2018, 113: 192-197.

6. **Diffusion and binding of Laponite clay nanoparticles into collagen fibers for the formation of leather matrix**

   Reprinted with permission from *Langmuir*, 2018, 34: 7379-7385. Copyright (2018) American Chemical Society.

7. Novel wet-white tanning approach based on Laponite clay nanoparticles for reduced formaldehyde release and improved physical performances

Reprinted with permission from *ACS Sustainable Chemistry & Engineering*, 2018, 108: 498-506. Copyright (2019) American Chemical Society.

Effects of Cr^{3+} on the Structure of Collagen Fiber

Bo Wu,[†,‡] Changdao Mu,[†] Guangzhao Zhang,[‡] and Wei Lin*[,†]

[†]*Department of Pharmaceutics and Bioengineering, National Engineering Laboratory for Clean Technology of Leather Manufacture, Sichuan University, Chengdu, Sichuan, China, and* [‡]*Hefei National Laboratory for Physical Sciences at Microscale, Department of Chemical Physics, University of Science and Technology of China, Hefei, Anhui, China*

Received May 3, 2009. Revised Manuscript Received June 13, 2009

We have investigated the effects of Cr^{3+} on the hierarchical structure of pigskin collagen fibers by use of scanning electron microscopy (SEM), X-ray photoelectron spectroscopy (XPS), wide-angle X-ray diffraction (WAXD), confocal laser micro-Raman spectroscopy (CLRS), and circular dichroism (CD). Our results demonstrate that the introduction of Cr^{3+} leads to the formation of a cluster of 20−40 nm between collagen fibrils, while the unique axial periodic structure (D periodicity) of the fibrils does not change. As the Cr^{3+} concentration increases, the order of intermolecular lateral packing, crystallite structure within helical chains, and N and C telopeptide regions decrease. The present study reveals that Cr^{3+} only cross-links with collagen but does not disrupt its triple helical structure.

Introduction

As the major structural components of the extracellular matrix, fibril-forming collagens provide the main mechanical support and structural organization of connective tissues such as skin, tendon, cartilage, and bone.[1] The specific functions are related to the hierarchical structure of collagen from triple helix to fiber. Actually, understanding the structure−function relations can not only help to reveal the underlying pathology of many diseases,[2,3] but also offer a powerful strategy for design and fabrication of new functionalized biomimetic materials.[4,5] It is well documented that the most abundant type I collagen comprises two identical α1(I) chains and a different α2(I) chain. Each of them has the repeating triplet amino acid sequence of Gly-X-Y, in which X and Y are frequently proline (Pro) and hydroxyproline (Hyp), respectively.[6] The three α-chains twist together into a unique triple helical structure. The collagen monomers (∼1.5 nm in diameter) consist of such a long helical region of ∼300 nm in length and short non-helical N and C-terminal telopeptides.[7] The quarter staggered arrangement of collagen molecules leads to collagen fibrils with a characteristic axial periodic structure, and the fibrils further assemble into a collagen fiber.[8−13]

Animal hides or skins contain dominantly type I collagen and are commercially important as natural frameworks largely utilized in medical, food, and leather industries. They are converted into leather upon tanning, and basic Cr^{3+} salts have been the most effective tanning agent.[14] It is recognized that Cr^{3+} ions cross-link with collagen so that the resultant leather has resistance to microbial attack and enhanced stability against wet and dry heat.[15] It is reported that Al^{3+} and Fe^{3+} ions can also form similar cross-links with collagen. In comparison with Cr^{3+} tanned leather, Al^{3+}-tanned leather has poorer endurance to water and high temperature, while Fe^{3+}-tanned leather is brittle with a poorer aging property. The strong binding of Cr^{3+} to collagen is thought to be responsible for the difference between them.[16] Similarly, because Ti^{4+} and Zr^{4+} ions form weak complexation with carboxyls in collagen, they can not act as good tanning agents.[15] Therefore, Cr^{3+} salts have been predominating among a variety of agents for more than 100 years. However, the interaction between Cr^{3+} and collagen at the molecular level and its effect on the hierarchical structure of collagen fiber remain largely unknown. It has been reported that Cr^{3+} ions can induce the supramolecular organization of isolated or hydrolyzed collagens in dilute solutions.[17,18] By use of synchrotron X-ray diffraction, Maxwell et al.[19,20] have studied the effect of liming and Cr^{3+} salt concentration on the collagen fibers and found that Cr^{3+} can alter the fibril packing of bovine collagen. Nevertheless, the effect of

---

*To whom correspondence should be addressed. E-mail: wlin@scu.edu.cn.
(1) Ricard-Blum, S.; Ruggiero, F.; Van der Rest, M. *Top. Curr. Chem.* **2005**, *247*, 35–84.
(2) Vitagliano, L.; Nemethy, G.; Zagari, A.; Scheraga, H. A. *J. Mol. Biol.* **1995**, *247*, 69–80.
(3) Orgel, J. P.; Miller, A.; Irving, T. C.; Fischetti, R. F.; Hammersley, A. P.; Wess, T. J. *Structure* **2001**, *9*, 1061–1069.
(4) Zhang, W.; Liao, S. S.; Cui, F. Z. *Chem. Mater.* **2003**, *15*, 3221–3226.
(5) Pires, M. M.; Chmielewski, J. *J. Am. Chem. Soc.* **2009**, *131*, 2706–2712.
(6) Piez, K. A.; Reddi, A. H. Molecular and aggregate structures of the collagens. In *Extracellular Matrix Biochemistry*; Piez, K. A., Reddi, A. H., Eds.; Elsevier: New York, 1984; pp 1–39.
(7) Hofmann, H.; Fietzek, P. P.; Kühn, K. *J. Mol. Biol.* **1980**, *141*, 293–314.
(8) Chapman, J. A.; Hulmes, D. J. S. Electron microscopy of the collagen fibril. In *Ultrastructure of the connective tissue matrix*; Ruggeri, A., Motta, P. M., Eds.; Martinus Nijhoff: Boston, 1984; pp 1–33.
(9) Chernoff, E. A. G.; Chernoff, D. *J. Vac. Sci. Technol. A* **1992**, *10*, 596–599.
(10) Baselt, D. R.; Revel, J. P.; Baldeschwieler, J. D. *Biophys. J.* **1993**, *65*, 2644–2655.
(11) Revenko, I.; Sommer, F.; Tran Minh, D.; Garrone, R.; Franc, J. M. *Biol. Cell* **1994**, *80*, 67–69.
(12) Gale, M.; Pollanen, M. S.; Markiewicz, P.; Goh, M. C. *Biophys. J.* **1995**, *68*, 2124–2128.

(13) Orgel, J. P.; Irving, T. C.; Miller, A.; Wess, T. J. *Proc. Natl. Acad. Sci. U.S. A.* **2006**, *103*, 9001–9005.
(14) Subramani, S.; Palanisamy, T.; Jonnalagadda, R. R.; Balachandran, U. N.; Thirumalachari, R. *Environ. Sci. Technol.* **2004**, *38*, 871–879.
(15) Covington, A. D. *Chem. Soc. Rev.* **1997**, *26*, 111–126.
(16) Gotsis, T.; Spiccia, L.; Montgomery, K. C. *J. Soc. Leather Technol. Chem.* **1992**, *76*, 195–200.
(17) Gayatri, R.; Sharma, A. K.; Rajaram, R.; Ramasami, T. *Biochem. Biophys. Res. Commun.* **2001**, *283*, 229–235.
(18) Lin, W.; Zhou, Y. S.; Zhao, Y.; Zhu, Q. S.; Wu, C. *Macromolecules* **2002**, *35*, 7407–7413.
(19) Maxwell, C. A.; Wess, T. J.; Kennedy, C. J. *Biomacromolecules* **2006**, *7*, 2321–2326.
(20) Maxwell, C. A.; Smiechowski, K.; Jan, A.; Sionkowska, A.; Wess, T. J *J. Am. Leather Chem. Assoc.* **2005**, *100*, 9–17.

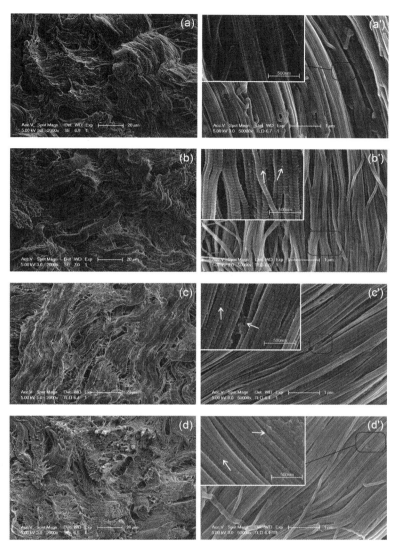

Figure 1. SEM microphotographs of collagen fibers treated with different concentrations of Cr^{3+}: (a) 0; (b) 25 μmol/g; (c) 125 μmol/g; (d) 225 μmol/g. (a′), (b′), (c′), and (d′) are the images for a, b, c, and d with a 50,000 higher magnification. The inset shows a zoom-in image.

Cr^{3+} ions on the helical backbone is not clear yet. It should be noted that Cr^{3+} is being disposed because of the environmental problem despite its unique leathering effect. Recently, some new combination tanning methods based on aluminum/silica/phosphonium or nano-SiO$_2$/oxazolidine have been developed.[21,22] However, such chrome-free leather exhibits poorer mechanical properties than those tanned by Cr^{3+}. Clearly, understanding the role of Cr^{3+} in the collagen structure is helpful to design alternative cross-linking agents which are environmental-friendly but as effective as or better than Cr^{3+} salts. Actually, transition metal ions play essential roles in folding, stability, biosynthesis, and transport of proteins, as well as the catalysis of biological process. Metal binding leads a protein to adopt a specific conformation so that it functions. The role of metal ions is important for understanding the phenomena in biology.[23,24]

In the present work, we have investigated the hierarchical structure of pigskin collagen in the presence of Cr^{3+} by use of field emission scanning electron microscopy (SEM), X-ray

---

(21) Luo, Z. Y.; Fan, H. J.; Liu, Y. S.; Li, H.; Peng, B. Y.; Shi, B. *J. Soc. Leather Technol. Chem.* **2008**, *92*, 252–257.
(22) Fathima, N. N.; Kumar, T. P.; Kumar, D. R.; Rao, J. R.; Nair, B. U. *J. Am. Leather Chem. Assoc.* **2006**, *101*, 58–65.

(23) Jackson, G. S.; Murray, I.; Hosszu, L. L.; Gibbs, N.; Waltho, J. P.; Clarke, A. R.; Collinge, J. *Proc. Natl. Acad. Sci. U.S.A.* **2001**, *98*, 8531–8535.
(24) Sigel, A.; Sigel, H.; Sigel, R. K. O. *Metal Ions in Life Sciences*; Wiley: Chichester, 2007.

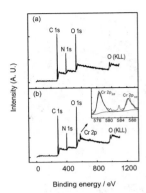

Figure 2. XPS survey spectra: (a) collagen fibers; (b) the collagen fibers treated with 275 μmol/g of Cr^{3+}. The inset shows the XPS Cr 2p core level spectra.

Table 1. Elemental Composition (atm %) Obtained by XPS from Collagen Fibers Treated with Different Concentrations of Cr^{3+}

$[Cr^{3+}]/(\mu mol/g)$	C(1s) %	N(1s) %	O(1s) %	Cr(2p) %
0	64.5 ± 1.5	14.0 ± 1.2	21.5 ± 0.4	0
25	62.9 ± 0.7	13.0 ± 0.4	23.1 ± 1.0	1.0 ± 0.1
75	63.6 ± 0.9	12.6 ± 0.4	22.5 ± 0.6	1.2 ± 0.0
125	63.5 ± 0.7	13.3 ± 0.5	21.7 ± 0.2	1.5 ± 0.1
175	63.5 ± 0.1	12.5 ± 0.1	22.5 ± 0.1	1.5 ± 0.1
225	62.9 ± 0.6	12.4 ± 0.2	23.3 ± 0.8	1.6 ± 0.2
275	62.5 ± 1.1	12.1 ± 0.1	23.2 ± 0.8	2.1 ± 0.4

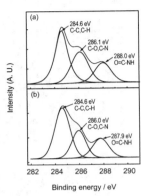

Figure 3. XPS C 1s core level spectra of collagen fibers treated with different concentrations of Cr^{3+}: (a) 25 μmol/g; (b) 275 μmol/g.

photoelectron spectroscopy (XPS), wide-angle X-ray diffraction (WAXD), confocal laser micro-Raman spectroscopy (CLRS), and circular dichroism (CD). Our aim is to explore Cr^{3+}-collagen interactions and understand the structure–property relations of collagen fiber.

Experimental Section

Sample Preparation. The raw collagen fibers were taken from the spinal part of fresh healthy pigskin. The skin was processed into leather under a condition for commercial leather-making. After mechanical dehairing and fleshing, and further defatting by surfactants, the skin was limed and delimed at room temperature to remove proteoglycan and other non-collagenous substances. The obtained pure collagen matrixes or skin substances were weighted as the base of dosage. Then the skin was pickled in a $NaCl/H_2SO_4$ solution at pH 2.5 for 12 h, and subsequently different amounts of chromic nitrate were added under stirring. Here, we only used chromic nitrate as the tanning agent so that the effect of anions can be neglected.[16,25] When the cross-section of the skin showed light blue-green, indicating that some Cr^{3+} ions penetrate into collagen fiber (~2 h), Na_2CO_3 was added to adjust the float pH to 4.0 under constant stirring. After an incubation of 24 h at 40 °C, the resultant leather was rinsed with distilled water to remove unfixed Cr^{3+}. The control samples used were only dehaired, fleshed, and defatted skin pieces which were originally taken from the antimere of those for Cr^{3+} treatment. The Cr^{3+} concentration ($[Cr^{3+}]$) here is defined as the moles of Cr^{3+} per gram of the skin substance. The frozen slices with a thickness of 30 μm were prepared using cryostat microtome (CM1900, Leica-Microsystems Corporation) at −35 °C. The slices were slowly air-dried at room temperature for 24 h before the measurements. Other chemicals were used as received.

SEM Observation. The morphologies of collagen fibers before and after Cr^{3+} salt treatment were observed by a field emission scanning electron microanalyser (JSM-6700, JEOL) operated at 5 KV. The slices were rinsed with methanol before sputter-coated with gold to avoid the possible contamination of the surface.

XPS Measurements. XPS measurements were performed on an X-ray photoelectron spectrometer (ESCALAB 250, Thermo-VG Scientific Corporation) with a monochromatic focused Al Kα X-ray source (1486.6 eV) to determine C, N, O, Cr, and other elements on the slice surface. High resolution core level spectra of Cr 2p, O 1s, N 1s, and C 1s were obtained using the monochromatic Mg X-ray source (1253.6 eV). The detection was performed at 45° with respect to the sample surfaces. The pressure in the sample chamber was maintained at 10^{-9} mbar. The peak assignment of elements was referred to NIST XPS Database, and the peak fitting was performed with the software package of XPS peak4.1 using a Shirley type background and Gaussian–Lorentzian peak shapes with Gaussian–Lorentzian ratio 80/20. Surface elemental stoichiometry was determined from peak-area ratio. For each sample, the analyses were made at three different points on the slices and atomic fractions were calculated from averages of the three.

WAXD Studies. Wide angle X-ray diffraction data were obtained using an 18 KW rotating anode X-ray diffractometer (MXPAHF, Japan) with a fixed Cu Kα radiation of 0.154 nm and a sample to detector distance of 10 cm. It allowed the wide angle diffraction feature of collagen fiber slice to be observed at a resolution of 0.08 nm. The sample was scanned in the range of diffraction angle 2θ from 5 to 60° with a scanning rate of 2°/min at ambient temperature and humidity. The Peakfit4 (AISL software), one-dimensional peak fitting program, was used to determine the peak size shapes and the integrated intensity of the linear profiles. The real lattice space d, that represents characteristic structure dimension of collagen fiber, can be calculated by $d = \lambda/(2 \sin \theta)$, where λ is the X-ray wavelength and θ is half of diffraction angle.[20]

CLRS Measurements. The slices of the collagen fibers with and without Cr^{3+} treatment were rinsed with methanol and air-dried just before CLRS measurement. The spectra were recorded using a confocal laser micro-Raman spectrometer (LabRam-HR, France JY Corporation) comprising a semiconductor laser operating at 785 nm in the range of 800–1800 cm^{-1}, with a collection time of 120 s and an accumulation of 3 scans at 1 cm^{-1} resolution. All the measurements were conducted at 25 °C. The spectra were corrected for biological

---

(25) Bienkiewicz, K. Tanning with inorganic agents. In *Physical Chemistry of Leather Making*; Bienkiewicz, K., Ed.; Robert E. Krieger Publishing Company: Malabar, FL, 1983; pp 324–384.

Table 2. Binding Energies and Atomic Fractions from the XPS C 1s Core Level Spectra for C Element in the Collagen Fibers Treated with Different Concentrations of Cr^{3+}

[Cr^{3+}]/(μmol/g)	C—C and C—H		C—N and C—O		O=C—NH	
	binding energy (eV)	atomic fractions (%)	binding energy (eV)	atomic fractions (%)	binding energy (eV)	atomic fractions (%)
0	284.6	55.1%	286.1	24.4%	288.0	20.5%
25	284.6	56.2%	286.1	22.8%	287.8	21.0%
75	284.6	58.2%	286.1	21.2%	288.0	20.6%
125	284.6	58.5%	286.1	21.3%	288.1	20.2%
175	284.6	57.6%	286.1	22.1%	287.9	20.4%
225	284.6	57.6%	286.0	22.4%	287.8	20.0%
275	284.6	59.4%	286.0	20.4%	287.9	20.2%

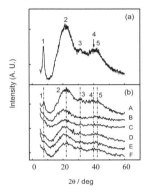

Figure 4. Wide angle X-ray diffraction linear intensity profile: (a) collagen fibers, (b) the collagen fibers treated with Cr^{3+}, where Cr^{3+} concentrations are (A) 25, (B) 75, (C) 125, (D) 175, (E) 225, and (F) 275 μmol/g, respectively.

fluorescence by subtraction of a fourth-order polynomial that was fitted to the spectrum according to reported procedures.[26] Spectral intensity was normalized using the CC vibrational mode of phenylalanine aromatic ring (1004 cm^{-1}) as an internal standard.

CD Measurements. The acid soluble collagen used for circular dichroism measurement was extracted from pigskin by 0.5 M acetic acid in the presence of pepsin. The details can be found elsewhere.[27] Collagen aqueous solutions were prepared by addition of chromic nitrate in acetate buffer (1 mM, pH 3.98) under stirring. Each solution was sealed and equilibrated for 18 h at room temperature. The final concentration of collagen was 2 × 10^{-4} g/mL. The CD spectra in the far UV region from 190 to 250 nm were recorded at 25 °C under nitrogen atmosphere on a J-810 CD spectropolarimeter (Jasco) using a quartz cell of 0.1 cm light path, with an average of 3 scans at a speed of 20 nm/min for each sample. A reference spectrum containing acetate buffer was also recorded. The resulting spectra were obtained after subtracting the reference spectrum and expressed in terms of molar ellipticity (E_m) at the wavelength λ in nm,[28]

$$E_m = \theta_\lambda / ncd \qquad \text{deg cm}^{-2}\,\text{dmol}^{-1} \qquad (1)$$

where θ_λ is the CD signal in mdeg, n is the number of amino acid residues in the protein chain, c is the molar concentration of the collagen solutions, and d is the path length in centimeter.

Results and Discussion

Figure 1 shows the morphologies of the collagen fibers. As expected, the interwoven flexuous fiber bundles span the entire image before Cr^{3+} is introduced. Namely, the collagen fiber has a macroscopically disordered network structure in nature. After the addition of Cr^{3+} salt, the collagen texture does not change. The thin uniform fibrils can be observed more clearly at a higher magnification (Figure 1a′-d′). They exhibit alternative dark (~0.4 D overlap) and light banding (~0.6 D gap) pattern, where the axial D periodicity is ~65 nm, in agreement with the reported value (~65.5 nm in skin).[19] It is known that only when the polypeptide chains of collagen are packed together in a definite and orderly manner and the chains maintain their native helical conformation can the banded structure be formed.[29] Therefore, the primary structure of the collagen is not disrupted by the added Cr^{3+}. We can also observe clusters with a diameter of 20-40 nm between collagen fibrils. Moreover, as Cr^{3+} concentrations increase, the clusters increase in number and are distributed more unevenly on fibrils.

Figure 2a shows that the XPS spectra of the collagen fiber only contains signals for the elements of C, N, and O. These elements together with hydrogen chemically constitute the amino acid residues of collagen.[30] After the introduction of Cr^{3+}, the bands at 577.2 and 586.9 eV can be observed (Figure 2b), which are attributed to asymmetric dual peaks of Cr $2p_{3/2}$ and $2p_{1/2}$ orbit, respectively.[31] Further deconvolution of Cr $2p_{3/2}$ peak leads to the peaks at 576.7 and 578.3 eV corresponding to the oxide (Cr_2O_3) and the hydroxide ($Cr(OH)_3$), respectively.[32] Electrons from the O (KLL) Auger energy level are also evidenced in each spectrum, but the amplitude of the corresponding peaks is low. The detailed atomic fractions for C, N, O, and Cr in various samples are given in Table 1. Therefore, the clusters observed by SEM are formed by Cr^{3+} ions cross-linking with the collagen segments around. The conversion of collagen into leather has been attributed to the Cr^{3+} mediated cross-linking, but the details are not clear yet.[15] Our results reveal that the formation of the clusters is responsible for the cross-linking between fibrils. This is maybe the reason that Cr^{3+} has unique tanning ability to turn collagen into leather with high hydrothermal stability.

Figure 3 shows that the collagen fibers before and after the treatment with Cr^{3+} have similar XPS C 1s core level spectra. The C 1s spectra can be deconvoluted into three components, which are attributed to carbon atom bound to carbon and hydrogen

---

(26) Buschman, H. P.; Deinum, G.; Motz, J. T.; Fitzmaurice, M.; Kramer, J. R.; Van der laarse, A.; Bruschke, A. V.; Feld, M. S. *Cardiovasc. Pathol.* **2001**, *10*, 69-82.
(27) Mu, C. D.; Li, D. F.; Lin, W.; Ding, Y. W.; Zhang, G. Z. *Biopolymers* **2007**, *86*, 282-287.
(28) Brown, E. M.; Farrell, H. M.; Wildermuth, R. J. *J. Protein Chem.* **2000**, *19*, 85-92.

(29) Highberger, J. H. *J. Am. Leather Chem. Assoc.* **1993**, *88*, 120-150.
(30) Sionkowska, A.; Wisniewski, M.; Kaczmarek, H.; Skopinska, J.; Chevallier, P.; Mantovani, D.; Lazare, S.; Tokarev, V. *Appl. Surf. Sci.* **2006**, *253*, 1970-1977.
(31) Ithurbide, A.; Frateur, I.; Galtayries, A.; Marcus, P. *Electrochim. Acta* **2007**, *53*, 1336-1345.
(32) Pradier, C. M.; Karmanl, F.; Telegdi, J.; Kalman, E.; Marcus, P. *J. Phys. Chem. B* **2003**, *107*, 6766-6773.

Table 3. WAXD Results for Collagen Fibers Treated with Different Concentrations of Cr^{3+} [a]

$[Cr^{3+}]/(\mu mol/g)$	intermolecular lateral packing (nm)	amorphous region (nm)	helical rise per residue (nm)	N telopeptide (nm)	C telopeptide (nm)
0	1.21	0.415	0.287	0.228	0.214
25	1.16	0.394	0.288	0.227	0.212
75	—	0.442	—	—	—
125	—	0.424	0.288	—	—
175	—	0.427	0.288	—	—
225	—	0.436	—	—	—
275	—	0.409	—	—	—

[a] — means that the peak is not distinct or is absent.

(C—C and C—H), carbon singly bound to nitrogen and oxygen (C—N and C—O), and carbon involved in an amide bond (O=C—NH), respectively. The binding energies and C atomic fractions at various chemical states derived from various collagen fibers are listed in Table 2. Note that the intensity of the component located at around 288.0 eV originating from the peptide bond[33] almost does not change as Cr^{3+} concentration increases, implying that the primary structure of collagen is not destroyed by Cr^{3+} binding.[34]

Figure 4a shows WAXD linear intensity profiles of the collagen fiber versus diffraction angle. From the position of the maxima in reciprocal space, the real lattice space values are obtained by Bragg's equation (Table 3). The peak 1 at 1.21 nm in real lattice space represents the characteristic intermolecular lateral packing within collagen fibrils.[35] The position of Bragg reflection at approximately 0.287 nm (peak 3) relates to the axial rise distance between the amino acid residues along collagen triple helices or helical rise per residue.[36] It is reported that the axial translation values for amino acid residues in the N and C telopeptide are 0.241 and 0.2 nm, respectively.[37] Obviously, the peak range of 0.25—0.208 nm in peak 4 and 5 agrees with the previous results. The two diffraction peaks partially overlap each other because of close position and similar intensity. The broad reflection at around 0.415 nm (peak 2) corresponds to amorphous scatter resulted from unordered components of collagen fiber. Collagen from bovine hide exhibits a similar WAXD pattern,[19] namely, homotypic collagens of mammals have similar structure.

Figure 4b shows the WAXD profiles of collagen fibers treated with different concentrations of Cr^{3+} salts. The structural information is given in Table 3. When $[Cr^{3+}] > 75\ \mu mol/g$, the peak of the intermolecular lateral packing almost disappears; namely, the level of the packing order decreases. This suggests that Cr^{3+} ions enter the space between collagen molecules within microfibrils. It is generally accepted that such intermolecular (inter-triple helix) cross-linking via Cr^{3+} complexes is critical for transformation of animal hides and skins into useful leather. Besides, the peak 3 corresponding to the helical rise per residue becomes weaker after Cr^{3+} is introduced. This is because Cr^{3+} may form complexation with side-chain carboxyls of aspartic and glutamic acids at a single helix chain or between chains within a triple helix,[38] leading to structural distortion of collagen molecules. However, the decrease in structural order does not correspond to a concomitant increase in the contribution of the amorphous region of the fiber. In other words, the degree of gelatinization is not increased. Therefore, Cr^{3+} does not destroy but distort the collagen helix.

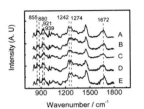

Figure 5. Raman spectra of collagen fibers treated with Cr^{3+} in the region of 800—1800 cm^{-1}, where Cr^{3+} concentrations are (A) 0, (B) 25, (C) 125, (D) 225, and (E) 275 $\mu mol/g$, respectively.

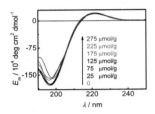

Figure 6. Far-UV CD spectra of the collagen solutions (2×10^{-4} g/mL) in the presence of Cr^{3+} ions.

Furthermore, as Cr^{3+} concentration increases, the diffraction intensity of N and C telopeptides decreases. It is believed that there are natural intermolecular cross-links situated between the non-helical terminal regions of adjacent or end-to-end connective collagen molecules, where the strongly polar residues are relatively rich and the packing of the polypeptide chains is of a lower degree of order.[37] As revealed by WAXD measurements above, the regions are favorable for the deposition of Cr^{3+} ions leading to the distortion of telopeptides. A similar phenomena has been observed in biomineralization with collagen as the matrix and apatite as mineral crystals.[39] The Cr^{3+} ions deposited between adjacent collagen helices can cross-link them and improve the strength of the collagen fibrils.

It is well-known that type I collagen has a special triple helix conformation, that is, right-handed superhelix formed by three left-handed helical strands. Its backbone structure and high proportion of imino acids can be well characterized by Raman spectra.[40,41] To further understand the effect of Cr^{3+}, we have investigated the collagen conformation by use of CLRS. Figure 5 shows a strong band at 1672 cm^{-1}, which is denoted as amide I band arising from the peptide carbonyl stretching vibration.

---

(33) Yokoyama, Y.; Kobayashi, T.; Iwaki, M. *Nucl. Instrum. Meth. B* **2006**, *24*, 237–240.
(34) Deng, S. B.; Bai, R. B. *Water Res.* **2004**, *38*, 2424–2432.
(35) Sionkowska, A.; Wisniewski, M.; Skopinska, J.; Kennedy, C. J.; Wess, T. J. *Biomaterials* **2004**, *25*, 795–801.
(36) Beck, K.; Brodsky, B. *J. Struct. Biol.* **1998**, *122*, 17–29.
(37) Orgel, J. P.; Wess, T. J.; Miller, A. *Structure* **2000**, *8*, 137–143.
(38) Covington, A. D.; Lampard, G. S.; Hancock, R. A.; Ioannidis, I. A.; Thisscos, St. Z. *J. Am. Leather Chem. Assoc.* **1998**, *93*, 107–120.

(39) Jiang, H. D.; Ramunno-Johnson, D.; Song, C. Y.; Amirbekian, B.; Kohmura, Y.; Nishino, Y.; Takahashi, Y.; Ishikawa, T.; Miao, J. W. *Phys. Rev. Lett.* **2008**, *100*, 038103.
(40) Frushour, B. G.; Koenig, J. L. *Biopolymers* **1975**, *14*, 379–391.
(41) Edwards, H. G. M; Farwell, D. W.; Holder, J. M.; Lawson, E. E. *J. Mol. Struct.* **1997**, *435*, 49–58.

Table 4. CD Data for Aqueous Collagen Solutions in 1.0 mM Acetate Buffer (pH 3.98) in the Presence of Cr^{3+}

$[Cr^{3+}]/(\mu mol/g)$	$\theta_{\lambda,\,max}$/nm	max $(E_m)/10^4$ deg cm^2 dmol^{-1}	$\theta_{\lambda,\,min}$/nm	min $(-E_m)/10^4$ deg cm^2 dmol^{-1}	R_{pn}
0	221	20.09	197	178.81	0.112
25	221	21.42	197	176.37	0.121
75	221	20.68	197	175.25	0.118
125	222	19.15	197	175.11	0.109
175	222	19.89	196	163.98	0.121
225	221	20.64	198	159.64	0.129
275	222	19.31	198	157.53	0.123

The amide III band originates from the N—H in plane deformation around 1274 cm^{-1} coupled to a C—N stretching mode at 1242 cm^{-1}, while peaks in the spectral region 1555—1575 cm^{-1} usually assigned to the amide II band are too weak to be observed in the Raman spectra. As shown in Figure 5, the introduction of Cr^{3+} does not alter the positions of amide I and III bands and the strong C—C stretch band of the peptide backbone around 939 cm^{-1}. Namely, the collagen after being treated with Cr^{3+} still has an intact triple helix structure.[42] Besides, the bands at 921 and 855 cm^{-1} corresponding to Pro and that at 880 cm^{-1} for Hyp are also observed because of strong Raman scattering of saturated side chain rings. Clearly, the helices retain Gly-X-Hyp and Gly-Pro-Y tripeptide sequence. The CLRS results further indicate that the introduction of Cr^{3+} does not destroy the triple helix of collagen, though it leads to the intra- and interchain cross-linking and the decrease of the order level of collagen fibers.

To examine Cr^{3+} ions induced conformational change of the collagen, we also investigated the collagen solutions in the presence of Cr^{3+} by use of CD (Figure 6). Normally, the denaturation of collagen or its triple helix structure destruction results in a red shift of the negative band at ~198 nm and disappearance of the positive band at ~221 nm.[43] The parameter R_{pn} or the ratio of positive peak intensity to the negative peak intensity is often used to describe the triple helical conformations for collagen and collagen-like peptides in solution.[44] When $[Cr^{3+}]$ < 175 μmol/g, the intensity and position of the band at 221 nm slightly change, so does the R_{pn} value (Table 4). The facts indicate the introduction of Cr^{3+} does not alter the helical conformation. This is consistent with previous results.[17,45] Gayatri et al. reported that Cr^{3+} ions do not change the helical conformation of collagen from rat tail tendon.[17] Fathima et al. have investigated the effect of Fe—tetrakis (hydroxymethyl) phosphonium (THP) on the conformation of type I collagen in acetate buffer by use of CD, and they revealed that the complex slightly alters the triple helical structure of collagen.[45] Further increasing Cr^{3+} concentration leads to a slight red shift of the negative band and a little increase of R_{pn}. This is likely due to the Cr^{3+}-collagen aggregates.

The combination of the above studies clearly indicate that Cr^{3+} clusters deposit among collagen fibrils, while smaller Cr^{3+} complexes diffuse into microfibrils, resulting in intermolecular and intramolecular cross-linking. The binding of Cr^{3+} to collagen leads to the decrease in the structural order, but it does not destroy the triple helices of collagen. Figure 7 illustrates the clustering of Cr^{3+} in collagen fibers.

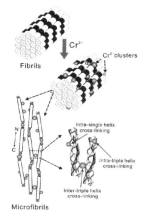

Figure 7. Schematic illustration showing the clustering of Cr^{3+} in the collagen fibers.

Conclusion

The studies on the structures of collagen fiber in pigskin treated by Cr^{3+} salt can lead to the following conclusions. Cr^{3+} ions and collagen segments around form a cluster of 20—40 nm between the fibrils acting as cross-linker. The cross-linking of Cr^{3+} does not destroy the triple helix conformation of collagen. As Cr^{3+} concentration increases, the order of intermolecular lateral packing and crystallite structure within helical chains decreases and the distortion of N and C telopeptide regions occurs because of the complexation between Cr^{3+} and the amino acid residues of collagen molecules. Obviously, a good tanning agent should form effective cross-linking with collagen but not destroy the backbone triple-helix. This is of significance for the development of new tanning agents in leather industry and new collagen based biomaterials.

Acknowledgment. The financial support of National Natural Science Foundation (NNSF) of China (20704028), Program for New Century Excellent Talents in University (NCET-06-0788), and Ministry of Science and Technology of China (2007CB936401) is gratefully acknowledged.

(42) Jastrzebskaa, M.; Zalewska-Rejdaka, J.; Wrzalikb, R.; Kocot, A.; Barwiński, B.; Mróz, I.; Cwalina, B *J. Mol. Struct.* **2005**, *744—747*, 789–795.
(43) Burjanadze, T. V.; Bezhitadze, M. O. *Biopolymers* **1992**, *32*, 951–956.
(44) Feng, Y.; Melacini, G.; Taulane, J. P.; Goodman, M. *J. Am. Chem. Soc.* **1996**, *118*, 10351–10364.
(45) Fathima, N. N.; Bose, M. C.; Rao, J. R.; Nair, B. U. *J. Inorg. Biochem.* **2006**, *100*, 1774–1780.

Fundamental Collagen Chemistry in Leather Making

Journal of Inorganic Biochemistry 117 (2012) 124–130

Contents lists available at SciVerse ScienceDirect

Journal of Inorganic Biochemistry

journal homepage: www.elsevier.com/locate/jinorgbio

Trivalent chromium and aluminum affect the thermostability and conformation of collagen very differently

Lirong He [a], Sumei Cai [a], Bo Wu [b], Changdao Mu [a], Guangzhao Zhang [b], Wei Lin [a,*]

[a] Key Laboratory of Leather Chemistry and Engineering of the Ministry of Education, National Engineering Laboratory for Clean Technology of Leather Manufacture, School of Chemical Engineering, Sichuan University, Chengdu, Sichuan, China
[b] Hefei National Laboratory for Physical Sciences at Microscale, Department of Chemical Physics, University of Science and Technology of China, Hefei, Anhui, China

ARTICLE INFO

Article history:
Received 20 April 2012
Received in revised form 27 August 2012
Accepted 28 August 2012
Available online 1 September 2012

Keywords:
Collagen
Thermostability
Conformation
US-DSC
AFM

ABSTRACT

Ultrasensitive differential scanning calorimetry (US-DSC) was used to directly measure the thermal transition temperature and energy change of acid soluble collagen in the presence of Cr^{3+} and Al^{3+} sulfates. The behavior of Cr^{3+} was analogous to kosmotropes in the cation Hofmeister series and increased the stability of collagen in dilute solutions. Meanwhile, the denaturational enthalpy change (ΔH) of collagen was substantially reduced with change to increasing Cr^{3+} concentration. This is likely due to the uni-point binding of Cr^{3+} with carboxyl groups of collagen side chains that could decrease the hydrogen-bonding in collagen and result in the increase of protein hydrophobicity. In the case of Al^{3+}, the interactions between the ions and collagen showed very different properties: at low and medium ion concentrations, the stability of the collagen was decreased; however, a further increase of Al^{3+} concentration led to a salting-out effect of collagen, indicating the Al^{3+} is a typical chaotropic ion. This striking difference of the two ions in the stabilization of collagen can be explained in terms of the different interactions between the cations and the carboxyl groups of collagen side chains. Additionally, we studied metal ion induced conformational change by the combination of circular dichroism (CD) and atomic force microscopy (AFM). CD measurements revealed that neither metal ion interactions of collagen with Cr^{3+} nor Al^{3+} ions destroyed the triple-helical backbone structure of collagen in the solution. AFM results further confirmed that the dehydration of collagen by Cr^{3+} is more significant than Al^{3+}, thus inducing the aggregation of collagen fibrils.

© 2012 Elsevier Inc. All rights reserved.

1. Introduction

Collagen is nowadays one of the most widely investigated proteins, because it not only represents the primary structural component of mammalian connective tissues including the skin, tendon, cartilage, bone, and ligament [1], but it also has wide industrial significance, as exemplified by the traditional leather industry and current biomedical applications [2–4]. A key procedure in the transformation of collagen fiber to leather is tanning, by which the characteristic change is the increased hydrothermal resistance, i.e., the rise in the denaturation temperature of native collagen [5]. It has long been recognized that among solo tannages, basic $Cr_2(SO_4)_3$ salt is unique in conferring high hydrothermal stability and desirable functional properties, whereas Al^{3+} sulfate tanned leather exhibits much poorer endurance to water and high temperature [6]. Therefore, Cr^{3+} salts have been dominant tanning agents since 1885, and the use of Al^{3+} salts has hitherto been extremely limited. Unfortunately, questions regarding the exact nature of the difference between Cr^{3+} and Al^{3+} in the stabilization of collagen have still not been satisfactorily answered [5]. It is a widely held view that the coordinate bonds between bi- or polynuclear chromium ions and ionized carboxyl groups of collagen side chains are more stable and responsible for the most effective tanning [7]. Al^{3+} does not form defined basic species and the weak interaction between Al^{3+} and collagen might be easier to break down on heating than that of Cr^{3+}-collagen. In the past, this was used to explain the considerable difference in hydrothermal stabilities of Cr^{3+} and Al^{3+} tanned leathers [6,8]. However, Covington et al. proposed that the metal tanning bonds did not break down during the thermal denaturation or heat shrinking of collagen fibers based on thermodynamic considerations [8] and ^{27}Al NMR tests [9]; but this still failed to explain the chemical and structural origins of the differences [5]. Recently, they attempted to establish a universal mechanism for further accounting the stabilization of collagen by using a link-lock model, in which the metal ions and counterions exhibit a synergistic effect in stabilizing the supramolecular water sheath around the collagen triple helix [10]. Although how the Cr^{3+} sulfate acts within the suprastructure remains unclear, the most important aspect of Covington's deliberations is to explore the role of water in the stabilization of collagen which has often been overlooked in the manufacture of leather [5]. In fact, many studies have indicated that the hydrothermal stability of unmodified and chemically modified collagen is dependent on the moisture content,

* Corresponding author. Fax: +86 28 85405237.
E-mail address: wlin@scu.edu.cn (W. Lin).

0162-0134/$ – see front matter © 2012 Elsevier Inc. All rights reserved.
http://dx.doi.org/10.1016/j.jinorgbio.2012.08.017

meaning that water is crucial for the collagen helix conformation [11–13].

Specific ion or Hofmeister effect has commonly been used to describe the interactions among ion, water and macromolecules [14]. Specific salt–protein interactions determine phenomena like protein folding, association, stability, and precipitation [15]. According to the ability of salts to precipitate certain proteins from an aqueous solution, the classical Hofmeister anion series is $SO_4^{2-} > CH_3COO^- > OH^- > F^- > Cl^- > Br^- \approx NO_3^- > I^- > ClO_4^- > SCN^-$. The species on the left of Cl^- are referred to as kosmotropes, which are strongly hydrated and have stabilizing and salting-out effects on proteins and macromolecules. In contrast, those on the right are called chaotropes, which are weakly hydrated and tend to act as protein denaturants and increase protein solubility [16,17]. As we know, the shrinkage temperatures of $Cr_2(SO_4)_3$, $CrCl_3$, $Cr(NO_3)_3$ and $Cr(ClO_4)_3$ tanned leathers are about 115, 95, 75, and 70 °C, respectively. The shrinkage temperature is reduced following the order of the so called Hofmeister series exactly, in agreement with the assumption that the anions influence the water envelope of the collagen [5]. It is worth noting that the cations can also be categorized as kosmotropes and chaotropes based on cation-specific effects though the latter are generally less pronounced in anion effects. The reason is due to the fact that anions have stronger interactions with water than cations of the same size and absolute charge density [18,19]. However, when direct ion–ion or ion–charged headgroup interactions are dominant, specific cation effects can be of the same order of magnitude as specific anion effects [19]. The typical ordering of cations and anions from kosmotrope to chaotrope is summarized in Fig. 1 [14].

Recently, by using ultrasensitive differential scanning calorimetry (US-DSC), we found that in dilute solution, the denaturation temperatures of acid soluble collagen in the presence of tanning salts including $Cr_2(SO_4)_3$ and $Al_2(SO_4)_3$ were not greatly enhanced as normally observed for collagen fibers, e.g., hide and skin. Hence the cation specific effects of Cr^{3+} and Al^{3+} may explain their difference in the stabilization of collagen, since the anion in the two salts is the same so that its effect can be neglected. In the present work, we have measured the thermal transition temperature and energy change of collagen under varied concentrations of Cr^{3+} and Al^{3+} by a high sensitivity microcalorimetric technique (0.02 μcal/s in precision). The conformational changes of the collagen have also been examined by circular dichroism (CD) and atomic force microscopy (AFM). This is apparently the first report of an US-DSC study of the interactions between Cr^{3+} and Al^{3+} sulfates with collagen at the molecular level, and explaining the results from the point of view of ion specificity.

2. Experimental section

2.1. Materials

The acid soluble collagen used in this study was extracted from the fresh adult bovine Achilles tendon in 0.5 M acetic acid with 2 w/w % pepsin to ensure the structural integrity of the collagen triple helix; the isoelectric point for the collagen is pH 7.4. The details can be found in our previous work [3]. $Cr_2(SO_4)_3 \cdot 6H_2O$ and $Al_2(SO_4)_3 \cdot 18H_2O$ were purchased from the Aladdin Reagent Company (Shanghai, China) and used without further purification. All chemical reagents used in this work were of analytical grade. The deionized water used in the experiment was purified with a UP Water Purification System (Ulupure, Shanghai, China) and had a minimum resistivity of 18.2 MΩ·cm.

2.2. Sample preparation

$Cr_2(SO_4)_3$, $Al_2(SO_4)_3$ and collagen solutions were freshly prepared with acetate buffer (12.4 mM sodium acetate, pH 4.0). Equal volumes of collagen solutions and corresponding salt solutions were mixed uniformly to get homogeneous solutions. The solutions were kept at pH 4 to ensure that the side chain carboxyl groups of aspartic and glutamic acid residues of collagen were mostly ionized [6]. The final concentration of collagen was 0.5 mg/mL for all the US-DSC measurements. For the CD and AFM tests, the collagen concentrations were even lower because of the limitations by the methodologies. The detailed solution concentrations were specified in the following respective measurements.

The overlap concentration (C^*) of the collagen was evaluated from $C^* = 3M/(4\pi N_A R_g^3)$ [20], with M, N_A and R_g being the molecular weight, the Avogadro constant, and the radius of gyration, respectively. Since the fibrillar Type I collagen can be regarded as a rod-like molecule [21], R_g was then estimated from $R_h/R_g = 3^{1/2}/(\ln(L/b) - \gamma)$ [22], where the hydrodynamic radius (R_h) is roughly equal to the molecular equivalent sphere radius (R) and evaluated from $M = 4/3\pi R^3 \rho$, and the rod length (L), rod diameter (b) and M for the Type I collagen are ~300 nm, ~1.5 nm and 30 kDa, respectively [23], and $\gamma \cong 0.3$ for the rod-like polymer [22]. Thus, the C^* value for the collagen used was estimated to be 44.6 mg/mL. Obviously, the collagen solutions (0.5 mg/mL at maximum) that we investigated in this work were in the dilute solution range.

2.3. Opacity measurements

The salt concentrations required for initiation of salting-out collagen were estimated by the sudden opacity increase of collagen solution [24]. The opacity of the collagen solutions at increasing salt concentrations by steps of 10 mM were measured at 430 nm using a UV-visible spectrophotometer (UV 2501PC, Shimadzu, Japan) at room temperature.

2.4. Ultra-sensitive differential scanning calorimeter measurements

Collagen solutions (0.5 mg/mL) with varied concentrations of Cr^{3+} or Al^{3+} sulfate were prepared in a sodium acetate buffer (12.4 mM, pH 4.0) and kept at room temperature for 24 h before the tests. The solutions were measured on a VP-DSC microcalorimeter (Microcal, Northampton, USA) with the matching acetate buffer (12.4 mM, pH 4.0) as the reference. The Cr^{3+} concentrations ranged from 0 to 1.5 mM, and the Al^{3+} concentrations from 0 to 380 mM until the opacity of collagen solutions was observed to increase. Baseline controls were obtained with corresponding acetate solutions in both sample and reference chambers, and subtracted from the sample runs. The sample solution and the reference solution were degassed for 30 min at ambient temperature (25 °C) before tests. All of the scans were

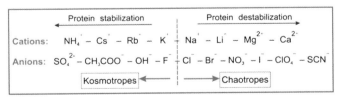

Fig. 1. Typical ordering of cations and anions from kosmotrope to chaotrope in Hofmeister series.

conducted from 20 to 60 °C at 1 °C/min. The calorimetric enthalpy change (ΔH) was calculated from the area under each peak. The melting temperature (T_m) was taken as the maximum usually centered at the transition [25,26]. The starting temperature (T_s) of the collagen denaturation was taken as the onset of the transition.

2.5. CD measurements

Collagen solutions (0.2 mg/mL) with varied concentrations of Cr^{3+} or Al^{3+} sulfate were prepared in a sodium acetate buffer (1 mM, pH 4.0). The final concentrations of Cr^{3+} were 0.08, 0.2, 0.4, and 0.6 mM, respectively. Al^{3+} concentrations were 4, 100, 120, 140, and 152 mM, respectively. All the solutions were sealed and equilibrated at room temperature for 24 h before tests. The CD spectra measurements in the far-UV region from 190 to 240 nm were carried out on a J-810 CD spectropolarimeter (Jasco) using a quartz cell of 0.1 cm light path. Each sample was scanned 3 times at a speed of 20 nm/min. A reference spectrum containing a matching acetate buffer was also recorded under the same condition as the baseline. The resulting spectra were obtained after subtracting the reference spectrum and expressed in terms of molar ellipticity (E_m) at the wavelength λ in nm [2], $E_m = \theta_\lambda / ncd$ deg cm^{-2}dmol^{-1}, where θ_λ is the CD signal in mdeg, n is the number of amino acid residues in the protein chain, c is the molar concentration of the collagen solutions, and d is the path length in centimeters.

2.6. AFM observations

The sample preparation for atomic force microscopy (AFM) observation was similar to our previous report [3]. A collagen solution (10 μL) at a concentration of 4 μg/mL was dropped onto a fresh mica substrate. Then a $Cr_2(SO_4)_3$ or $Al_2(SO_4)_3$ solution (10 μL) in pH 4.0 sulfuric acid at designated concentrations (0.033, 0.167, 0.333 μg/mL, respectively) was carefully added onto it. The final ratios of Cr^{3+}/collagen or Al^{3+}/collagen were 25, 125, and 250 μmol/g, respectively. The samples were dried in a desiccator for 24 h at room temperature before testing. AFM observations were then performed in a non-contact (tapping) mode on a Dimension 3100 Nanoscope IV with Silicon TESP cantilevers (SPM-9600, Shimadzu, Japan). For each sample, the tests were made at least three times to confirm the consistency of the observed morphology.

3. Results and discussion

3.1. Thermodynamic characteristics of acid soluble collagen in the presence of Cr^{3+}

Fig. 2 shows the temperature dependence of the specific heat capacity (C_p) of the Type I collagen in a pH 4 acetate buffer in the presence of $Cr_2(SO_4)_3$ with different concentrations. The typical US-DSC curve for the acid soluble collagen is similar to that in our previous studies [27]. It is characterized by two endothermic peaks with a pre-transition at $T_{m1} \sim 36$ °C and a main denaturational transition at $T_{m2} \sim 41$ °C. Conventionally, the shrinkage temperature for chromed leather or thermal transition temperature for Cr^{3+}-crosslinked collagen fiber shows increases with the presence of Cr^{3+} salt [2,28]. However, in this work, both T_m positions for the bimodal transition are not apparently shifted to higher temperatures with the increase of Cr^{3+} concentrations as expected. The starting temperature of the main transition (T_{s2}), however, is increased by about 1 °C (Fig. 3a) indicating the improved stability of the collagen by the introduction of Cr^{3+} salts. This likely suggests that the uni-point binding of one Cr^{3+} ion with one carboxyl group of collagen can occur dominantly in such dilute solutions [29], since it has been generally accepted that the reaction which determines the high hydrothermal stability is multi-point crosslinking, whereas uni-point fixation probably provides little hydrothermal

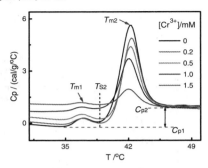

Fig. 2. Temperature dependence of the specific heat capacity (C_p) of the Type I collagen in a pH 4 acetate buffer in the presence of Cr^{3+} sulfate at different concentrations.

stability [6]. Note that a weak linear increase of the onset temperature in main transition, i.e. T_{s2}, with the Cr^{3+} concentration suggests the kosmotropic properties of Cr^{3+}.

It also can be seen clearly from Fig. 2 that the two transition peaks decrease distinctly with the addition of Cr^{3+} salt, and the corresponding enthalpy changes (ΔH_1 and ΔH_2) are shown in Fig. 3b and c, respectively. The calorimetric enthalpy ΔH of the transition denotes the energy required for the destruction of the hydrogen bonds which maintain the collagen triple helixes [24]. The regular decrease of the ΔH_1 and

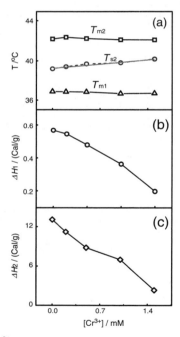

Fig. 3. The Cr^{3+} concentration dependence of the transition temperatures (T_{m1}, T_{m2}), the starting temperature of the main transition (T_{s2}) (a), and the enthalpy change of the pretransition (ΔH_1) (b) and main transition (ΔH_2) (c). Symbols denote experimental points, and solid lines serve to guide the eye.

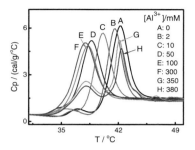

Fig. 4. Temperature dependence of the heat capacity (C_p) of the Type I collagen in a pH 4 acetate buffer in the presence of Al^{3+} sulfate at different concentrations.

ΔH_2 with the Cr^{3+} content indicates that partial hydrogen bonds of collagen are destroyed owing to the binding of Cr^{3+}, especially along the polypeptide backbone. It is worthy to note that the heat capacity of the collagen before thermal denaturation (C_{p1}) increases with the Cr^{3+} concentration, whereas the one corresponding to the complete unfolded state (C_{p2}) remains the same, revealing the Cr^{3+} concentration dependence of the collagen conformation. It is known that the C_p represents the hydrophobicity of the macromolecules; namely, a high C_p means hydrophobicity, whereas a low value indicates hydrophilicity [30–33]. Here the displacement of bound water involved in hydrogen bonds by Cr^{3+} ions might be responsible for the increased hydrophobicity of the collagen. Furthermore, the conceivable displacement process would certainly reduce the hydrogen bonds, which in turn leads to the decrease of the denaturational enthalpy change. Obviously, the more the Cr^{3+} ions bind to the collagen molecules, the lower is the ΔH.

3.2. Thermodynamic characteristics of acid soluble collagen in the presence of Al^{3+}

The effect of $Al_2(SO_4)_3$ on the thermal stability of collagen in a pH 4 acetate buffer has also been examined by US-DSC (Fig. 4). Unlike Cr^{3+} cations, three concentration ranges for Al^{3+} can be clearly distinguished according to the dependence of the collagen thermal stability on the ion concentration. The Al^{3+} ion overall exhibits a typical chaotropic ion effect similar to the reported mono- or bivalent salts [24]. At Al^{3+} ion concentrations below 50 mM, both the pre-transition and major transition shift to lower temperatures in parallel (Fig. 5a), indicating the reduction in collagen stability. This is generally attributed to the screening of the electrostatic interactions between the charged amino acid residues of collagen by salt [24]. At an intermediate salt concentration (50 mM<[Al^{3+}]<300 mM), the pre-transition merges into the main transition peak (Fig. 4), and the temperature of collagen unfolding (T_{m2}) is further reduced. Generally, specific ion effects are often measured at intermediate concentrations (e.g., above 0.1 M), where the long range electrostatic interactions are significantly masked, and the short-range ion-specific interactions can be observed [14,34]. Different salts either slightly stabilize or further destabilize the protein according to their position in the Hofmeister series. Similarly, the chaotropic ion effect of increasing Al^{3+} concentration on decreasing collagen stability is associated with indirect protein–salt interactions exerted via competition for water molecules between ions and the protein surface, which is reflected by the constant C_{p1}. The mechanism will be discussed in greater detail below.

At a higher Al^{3+} concentration, an abrupt increase of the T_m is observed (Fig. 5a), which is accompanied by a sudden increase of the solution opacity (data not show), indicating the occurrence of partial salting-out and aggregation of collagen. The enhanced stability of aggregated collagen in high Al^{3+} salts is consistent with the measured higher denaturation temperature of collagen fibers in comparison to the dissolved collagen [2,35], because the thermal stability of biological macromolecules is substantially influenced by their level of hydration [36]. It is interesting to note that the main transition splits into two peaks at the Al^{3+} concentrations above 300 mM (Fig. 4). Such splitting was also observed in a report for other sulfate salts; this splitting was proposed to result from specific interactions of sulfate anions with collagen molecules [24]. However, we speculate that it is due to partial aggregation occurring in the solutions. In addition,

Table 1
The minimum concentration of Cr^{3+} and Al^{3+} sulfates required to precipitate collagen from a 0.5 mg/mL solution in acetate buffer (pH 4) as estimated by opacity measurements.

Salt	Metal ion concentration [M^{3+}]/(mM)	Oxide content M_2O_3 (% w/w collagen)
$Cr_2(SO_4)_3$	2	30.4
$Al_2(SO_4)_3$	350	4080

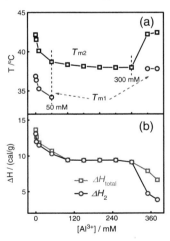

Fig. 5. Al^{3+} concentration dependence of the transition temperatures (T_{m1}, T_{m2}) (a), and the enthalpy change (ΔH_{total}, ΔH_2) (b). Symbols denote experimental points, and lines serve to guide the eye.

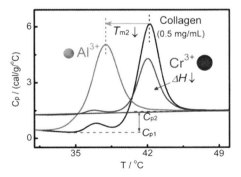

Fig. 6. Temperature dependence of the heat capacity (C_p) of the Type I collagen in a pH 4 acetate buffer in the presence of 300 mM Al^{3+} (green line) and 1.5 mM Cr^{3+} (blue line). The US-DSC thermogram showing the different effects of Cr^{3+} and Al^{3+} on the stability of an acid soluble collagen. (For interpretation of the references to color in this figure legend, the reader is referred to the web version of this article.)

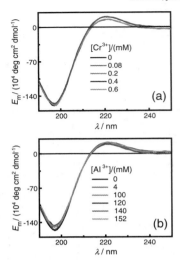

the activation entropy that is calculated from the total enthalpy changes (ΔH_{total}) is reduced in the range of $[Al^{3+}] > 300$ mM (Fig. 5b), which agrees with the reported calorimetric measurement results for collagen in a quasi-solution state [36].

3.3. Mechanism of Cr^{3+} and Al^{3+} in the stabilization of collagen

By comparing the collagen salting-out concentration, one can see the great difference between Cr^{3+} and Al^{3+} in the stabilization of collagen, as shown in Table 1. The collagen solution with a concentration of 0.5 mg/mL is salted out by as low as 2 mM Cr^{3+} (equal to 30.4 w/w % Cr_2O_3/dry collagen), but it remains dissolved in an $Al_2(SO_4)_3$ solution until the Al^{3+} exceeds 300 mM (~4080 w/w % Al_2O_3/dry collagen). Obviously, the salting-out effect of Cr^{3+} is two orders of magnitude stronger than that of Al^{3+}. Because of this large difference in salting out effects by the two metal ions, in accordance with the Hofmeister classification, Cr^{3+} can be considered as a kosmotropic cation which enhances the protein stability and induces salting-out, whereas Al^{3+} is a typical chaotropic ion which decreases the protein stability and induces salting-in. At very high ion concentrations, even chaotropes can induce salting-out [24]. Note that such large difference for the two trivalent ions must imply the different mechanisms that affect the thermal stability of collagen.

According to Collins' concept of matching water affinities, only oppositely charged ions of equal water affinity can form strong ion pairs, leading to the strong binding of the counterions onto the oppositely charged headgroups on the macromolecular surface [37]. Since

Fig. 7. Far-UV CD spectra of the collagen solutions (0.2 mg/mL) in the presence of Cr^{3+} (a) and Al^{3+} ions (b).

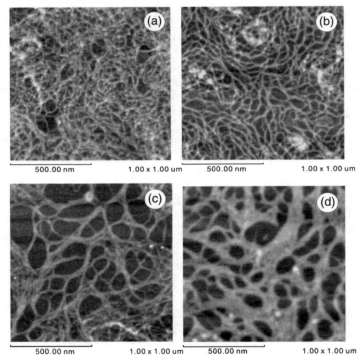

Fig. 8. AFM images of the Cr^{3+}-collagen with different ratios of Cr^{3+}/collagen: (a) 0; (b) 25 μmol/g; (c) 125 μmol/g; (d) 250 μmol/g, respectively.

the alkyl carboxylate is a hard (small and strongly hydrated) anionic headgroup which can be classified as a kosmotrope in a sort of Hofmeister series for charged headgroups, it prefers to form direct ion pairs with smaller cations located on the right of the cation Hofmeister series [38]. In other words, the ionized carboxyl groups available on the collagen surface prefer smaller Al^{3+} (0.5 Å) than relatively larger Cr^{3+} (0.69 Å) cations [39]. This is in agreement with the reported phenomenon that Al^{3+} can react quickly with collagen carboxyls possibly in a loose association that is entropy favored. Thus chrome tanning can be accelerated by pretreatment with aluminum; in this process, loosely bound Al^{3+} ions are displaced by Cr^{3+}, aluminum acting as a catalyst [6]. In addition, the empirical law of matching water affinities also states qualitatively that the interaction of proteins containing many anionic carboxylates with kosmotropes is more pronounced than with chaotropes [14]. Indeed, as shown by the US-DSC thermogram in Fig. 6, our measurements indicate that the ability of Al^{3+} to decrease collagen stability (lowering of T_m by ~4 °C) is more pronounced than the ability of Cr^{3+} to increase collagen stability (increasing of T_s by ~1 °C). Therefore, the concept of matching water affinities [37], together with the recently developed Hofmeister-like series for headgroups [37], can explain at least qualitatively the difference between Cr^{3+} and Al^{3+} in affecting the thermal stability of collagen in a dilute solution. Note that the effect of Al^{3+} on destabilizing protein may be a contributing factor to Al-triggered brain diseases [40,41]. The findings on the thermodynamics of the interaction of Al^{3+} with collagen provide potential relevance to aluminum toxicity.

3.4. Conformational changes of acid soluble collagen induced by Cr^{3+} and Al^{3+}

To examine the ion specific effects on the conformational change of the collagen, we also investigated the collagen solutions in the presence of Cr^{3+} or Al^{3+} by using CD (Fig. 7). It has been proven that complete denaturation of the collagen or its triple helix destruction usually results in a red shift of the negative band at ~198 nm and disappearance of the positive band at ~221 nm, characteristics of a collagen triple helix [42]. Furthermore partially denatured collagen gives a CD spectrum with lower intensity (especially at ~221 nm) as well as a lower value of R_{pn}, the ratio of the positive peak intensity to the negative peak intensity [27]. In contrast, the increase in the R_{pn} indicates the aggregation of collagen molecules [2,43]. Fig. 7 shows that addition of Cr^{3+} or Al^{3+} to the collagen solution did not alter the CD spectra significantly at the similar weight ratio of ions to collagen compared to US-DSC measurement. This information indicates that the interactions between the cations and collagen may occur dominantly on the accessible protein surface, and therefore the backbone structure of the collagen is not altered. At a very high Al^{3+} concentration, a slight increase of the R_{pn} from 0.13 to 0.16 is likely due to the formation of Al^{3+}-collagen aggregates by salting-out, which agrees with the above US-DSC results. However, at a relatively higher Cr^{3+} concentration, the negligible decrease of R_{pn} suggests some conformational distortion. As discussed before, the US-DSC results reflect partial breakage of the hydrogen bonds induced by Cr^{3+} binding. Here, the maintained triple helix conformation implies that the Cr^{3+}

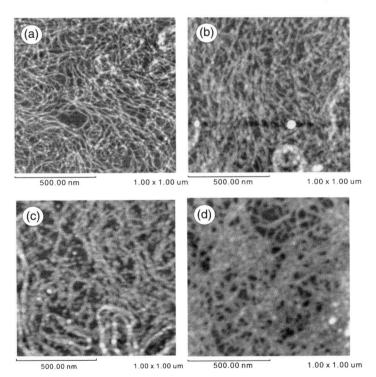

Fig. 9. AFM images of the Al^{3+}-collagen with different ratios of Al^{3+}/collagen: (a) 0; (b) 25 μmol/g; (c) 125 μmol/g; (d) 250 μmol/g, respectively.

does not destroy the triple helix conformation of collagen, but stabilizes the molecules by displacement of the bound water around them[44]. The results coincide with those reported for collagen fibers treated with Cr^{3+} nitrate[2].

Figs. 8 and 9 show the AFM morphological changes of collagen induced by Cr^{3+} and Al^{3+}, respectively. The pure Type I collagen in sulfuric solutions at pH 4.8 (Figs. 8a and 9a) exhibits a typical fibrillar structure on the mica substrate where many curved molecules or microfibrils are overlapped with one another, similar to our previous reports at the pH 2.5–3.0 in acetic acid solution[3,45]. In contrast, at given lower Cr^{3+} and Al^{3+} concentrations (Figs. 8b and 9b), a similar slight increase in the fibril diameter can be observed for both, suggesting the nonspecific electrostatic interactions of the cations and collagen. On further increase of the ion concentrations, the collagen with Cr^{3+} exhibits meshy fibrous network topographies and conformational heterogeneity (Fig. 8c and d), indicating the occurrence of aggregation or salting out effect; whereas in the presence of Al^{3+}, such aggregation is not significant (Fig. 9c and d). In repeated experiments, the Al^{3+} always shows much weaker astringent or dehydration on collagen fibrils than Cr^{3+}, supporting US-DSC results and empirical results of tanning in the manufacture of leather.

4. Conclusions

The studies on the thermal stability of the fibrillar Type I collagen by US-DSC as well as its conformational change in the presence of Cr^{3+} and Al^{3+} sulfates can lead to the following conclusions. The Cr^{3+} has a strong salting-out effect on collagen, analogous to the kosmotropes in the Hofmeister series which increase protein stability. The binding of Cr^{3+} with collagen leads to the decrease of collagen hydrogen-bonding and the increase of the protein hydrophobicity; in contrast, the Al^{3+} exhibits a typical chaotropic effect, which decreases collagen stability and induces salting-in at lower and medium concentrations. At very high concentrations, however, Al^{3+} induces salting-out. The apparent deviation of the two trivalent metal ions in the stabilization of collagen can be explained by the different interactions between the cations and the carboxyl groups of collagen side chains. Such interactions though different do not alter the backbone triple-helical structure of collagen as revealed by CD measurements. AFM results further indicate that the collagen dehydration by Cr^{3+} is more significant than Al^{3+}, thus inducing the aggregation of the fibrils. The present study reveals that the ion specificity is attributed to surface-related ion-charged headgroup interactions, and also provides insight into tanning mechanisms at the molecular level.

Abbreviations

AFM	atomic force microscopy
C*	overlap concentration
CD	circular dichroism
C_p	heat capacity
US-DSC	ultrasensitive differential scanning calorimetry
E_m	molar ellipticity
R_{pn}	ratio of the positive peak intensity to the negative peak intensity
T_m	melting temperature
T_s	starting temperature
ΔH	enthalpy change

Acknowledgments

The financial support of the National Natural Science Foundation (NNSF) of China (21074074, 21176159), and the Science and Technology Planning Project of Sichuan Province (2011HH0008) is gratefully acknowledged.

References

[1] C.H. Lee, A. Singla, Y. Lee, Int. J. Pharm. 221 (2001) 1–22.
[2] B. Wu, C.D. Mu, G.Z. Zhang, W. Lin, Langmuir 25 (2009) 11905–11910.
[3] L.R. He, C.D. Mu, J.B. Shi, Q. Zhang, B. Shi, W. Lin, Int. J. Biol. Macromol. 48 (2011) 354–359.
[4] K.H. Gustavson, The Chemistry and Reactivity of Collagen, Academic Press, New York, 1956.
[5] G. Reich, From Collagen to Leather–the Theoretical Background, BASF Service Center Media and Communications, Ludwigshafen, in: G. Reich (Ed.), 2007, pp. 98–124.
[6] A.D. Covington, Chem. Soc. Rev. 26 (1997) 111–126.
[7] E.M. Brown, M.M. Taylor, J. Am. Leather Chem. Assoc. 98 (2003) 408–414.
[8] A.D. Covington, G.S. Lampard, R.A. Hancock, I.A. Ioannidis, J. Am. Leather Chem. Assoc. 93 (1998) 107–120.
[9] A.D. Covington, R.A. Hancock, I.A. Ioannidis, J. Soc. Leather Technol. Chem. 73 (1989) 1–8.
[10] A.D. Covington, L. Song, O. Suparno, H.E.C. Koon, M.J. Collins, J. Soc. Leather Technol. Chem. 92 (2008) 1–7.
[11] C.A. Miles, N.C. Avery, V.V. Rodin, A.J. Bailey, J. Mol. Biol. 346 (2005) 551–556.
[12] A. Rochdi, L. Foucat, J. Renou, Biopolymers 50 (1999) 690–696.
[13] J. Bella, B. Brodsky, H.M. Berman, Structure 3 (1995) 893–906.
[14] W. Kunz, Curr. Opin. Colloid Interface Sci. 15 (2010) 34–39.
[15] P.L. Nostro, B.W. Ninham, Chem. Rev. 112 (2012) 2286–2322.
[16] Y. Zhang, P.S. Cremer, Curr. Opin. Chem. Biol. 10 (2006) 658–663.
[17] X. Chen, S.C. Flores, S.M. Lim, Y. Zhang, T. Yang, J. Kherb, P.S. Cremer, Langmuir 26 (2010) 16447–16454.
[18] I. Shechter, O. Ramon, I. Portnaya, Y. Paz, Y.D. Livney, Macromolecules 43 (2010) 480–487.
[19] W. Kunz, R. Neueder, in: W. Kunz (Ed.), Specific Ion Effects, World Scientific Publishing, Singapore, 2010, p. 35.
[20] Y.W. Ding, G.Z. Zhang, J. Phys. Chem. C 111 (2007) 5309–5312.
[21] E.A.G. Chernoff, D.A. Chernoff, J. Vac. Sci. Technol. A 10 (1992) 596–599.
[22] I. Teraoka, in: I. Teraoka (Ed.), Polymer Solutions: an Introduction to Physical Properties, New York, 2002, p. 187.
[23] G. Penners, Z. Priel, A. Silberberg, J. Colloid Interface Sci. 80 (1981) 437–444.
[24] R. Komsa-Penkova, R. Koynova, G. Kostov, B.G. Tenchov, BBA- Protein Struct. Mol. 1297 (1996) 171–182.
[25] H. Chen, Q. Zhang, J. Li, Y.W. Ding, G.Z. Zhang, C. Wu, Macromolecules 38 (2005) 8045–8050.
[26] X. Wang, X. Qiu, C. Wu, Macromolecules 31 (1998) 2972–2976.
[27] C.D. Mu, D.F. Li, W. Lin, Y.W. Ding, G.Z. Zhang, Biopolymers 86 (2007) 282–287.
[28] C.A. Maxwell, K. Smiechowski, Z. Jan, A. Sionkowska, T.J. Wess, J. Am. Leather Chem. Assoc. 100 (2005) 9–17.
[29] W. Lin, Y.S. Zhou, Y. Zhao, Q.S. Zhu, C. Wu, Macromolecules 35 (2002) 7407–7413.
[30] B. Madan, K. Sharp, J. Phys. Chem. 100 (1996) 7713–7721.
[31] K.A. Sharp, B. Madan, J. Phys. Chem. B 101 (1997) 4343–4348.
[32] N.V. Prabhu, K.A. Sharp, Annu. Rev. Phys. Chem. 56 (2005) 521–548.
[33] A. Cooper, Biophys. Chem. 85 (2000) 25–39.
[34] Z. Ao, G.M. Liu, G.Z. Zhang, J. Phys. Chem. C 115 (2011) 2284–2289.
[35] T. Gotsis, L. Spiccia, K.C. Montgomery, J. Soc. Leather Technol. Chem. 76 (1992) 195–200.
[36] C.A. Miles, M. Ghelashvili, Biophys. J. 76 (1999) 3243–3252.
[37] K.D. Collins, Methods 34 (2004) 300–311.
[38] N. Vlachy, B. Jagoda-Cwiklik, R. Vha, D. Touraud, P. Jungwirth, W. Kunz, Adv. Colloid Polym. Sci. 146 (2009) 42–47.
[39] R.D. Shannon, Acta Cryst. A 32 (1976) 751–767.
[40] J. Lemire, V.D. Appanna, J. Inorg. Biochem. 105 (2011) 1513–1517.
[41] S. Tang, R. MarcColl, P.J. Parsons, J. Inorg. Biochem. 60 (1995) 175–185.
[42] R. Usha, T. Ramasami, Thermochim. Acta 409 (2004) 201–206.
[43] R. Gayatri, A.K. Sharma, R. Rajaram, T. Ramasami, Biochem. Biophys. Res. Commun. 283 (2001) 229–235.
[44] I.G. Mogilner, G. Ruderman, J.R. Grigera, Collagen stability, hydration and native state, J. Mol. Graph. Model. 21 (2002) 209–213.
[45] D.F. Li, C.D. Mu, S.M. Cai, W. Lin, Ultrason. Sonochem. 16 (2009) 605–609.

Appendix III

Journal of Cleaner Production 139 (2016) 1512–1519

Contents lists available at ScienceDirect

Journal of Cleaner Production

journal homepage: www.elsevier.com/locate/jclepro

A comprehensive evaluation of physical and environmental performances for wet-white leather manufacture

Jiabo Shi [a], Rita Puig [b, *], Jun Sang [a], Wei Lin [a, **]

[a] *National Engineering Laboratory for Clean Technology of Leather Manufacture, Key Laboratory of Leather Chemistry and Engineering of Ministry of Education, Sichuan University, Chengdu, 610065, China*
[b] *Universitat Politècnica de Catalunya, Igualada School of Engineering (EEI), Igualada, Barcelona, Spain*

ARTICLE INFO

Article history:
Received 2 March 2016
Received in revised form
22 August 2016
Accepted 23 August 2016
Available online 25 August 2016

Keywords:
Wet-white leather
Tannic acid
Laponite nanoclay
Combination tannage
Life cycle assessment

ABSTRACT

This paper presents the comprehensive evaluation results of physical and environmental performances for a novel wet-white (chrome-free) leather manufacturing. The tanning process is optimized as 15 wt% tannic acid (TA) combination with 4 wt% Laponite nanoclay, giving the leather with shrinkage temperature (T_s) above 86 °C. Inductively coupled plasma-atomic emission spectrometry (ICP-AES) measurements indicate that Laponite can be evenly and tightly bound within the leather matrix, which is further confirmed by scanning electron microscopy and energy dispersive X-ray (SEM-EDX) spectroscopy analysis. The resultant wet-white leathers have reasonable good physical properties that can meet the standard requirements for furniture leather without containing hazardous Cr(VI) and formaldehyde. Further life cycle assessment (LCA) studies shows that tanning process is the main contributor to environmental impact categories in the wet-white tanning process, and tannic acid is the most significant substance factor. Compared to conventional chrome tanning, the wet-white tanning process exhibits much lower abiotic depletion potential (ADP), and reduced global warming potential (GWP) and human toxicity potential (HTP) impacts due to the nature of vegetable tanning; whereas, GWP excluding biogenic carbon and energy consumption are higher owing to prolonged run time.

© 2016 Elsevier Ltd. All rights reserved.

1. Introduction

In recent decades, increasing issues on the environmental and human health risks have greatly accelerated people's focus on developing more environmental-friendly leather chemicals (Krishnamoorthy et al., 2013) and cleaner technologies for leather manufacturing (Wang et al., 2016). As the decisive procedure during the structural transformation of collagen fibers into leather matrix, tanning process has been widely recognized and studied (Covington, 1997). Its main environmental burdens are originated from the consumptions of tanning materials and the emissions of solid wastes and wastewaters (Zhang et al., 2016).

Currently, chrome tanned leather shares a high proportion of all leather articles worldwide (Covington, 2009). Some of chemicals used in post-tanning procedures can lead to the possible conversions of Cr(III) to Cr(VI) (Chandra Babu et al., 2005). The formation of Cr(VI) can cause severe allergic contact dermatitis in human skins and can elicit dermatitis at very low concentrations (Hansen et al., 2003). In 2014, a more stringent restriction on the Cr(VI) concentration of leather articles was revised by the European Commission in Annex XVII of regulations of Registration, Evaluation, Authorization and Restriction of Chemicals (REACH). In the new restrictions, leather articles, coming into direct and prolonged or repetitive contact with the skin, should not be placed on the market if they contain Cr(VI) in concentrations equal to or higher than 3 mg/kg. The restriction and risk of Cr(VI) have directly affected the production, consumption and circulation of chrome tanned leather articles. Nowadays China is the largest country of leather production in the world. As a dominant traditional manufacture, leather-making industry in China is now faced with counteraction of environmental pollution and green trade barrier in leather exports.

In the case, wet-white (chrome-free) tanning towards Ecoleather manufacture has been increasingly emphasized in order to replace or reduce conventional chrome tanning, like the novel tannage based on nano-SiO$_2$ (Liu et al., 2010), and titanium based tannage (Crudu et al., 2014), etc. Among these, the combination

* Corresponding author. Tel.: +34 93 803 53 00.
** Corresponding author. Tel.: +86 28 8546 0819.
E-mail addresses: rita.puig@eei.upc.edu (R. Puig), wlin@scu.edu.cn (W. Lin).

http://dx.doi.org/10.1016/j.jclepro.2016.08.120
0959-6526/© 2016 Elsevier Ltd. All rights reserved.

Fundamental Collagen Chemistry in Leather Making

tannage of vegetable tannin and nano-silicate for wet-white leather manufacture (Shi et al., 2013) has been considered as a promising alternative to conventional chrome tannage due to natural resources and appropriate tanning properties. The developed wet-white tanning system is completely free of chromium and formaldehyde.

On the other hand, environmental assessments of a new tanning process related to human health and environmental risks are also essential for leather manufacturing. Various environmental assessment techniques are involved including materials flow analysis (Brunner and Rechberger, 2004), environmental impact assessment (Jay et al., 2007), multi-objective optimization (Erol and Thöming, 2005) and human health and environmental risk assessment (ERA) (Bridges, 2003). However, these assessment techniques are generally not comprehensive enough (Burgess and Brennan, 2001), and ERA is often more a legal procedure than a detailed environmental assessment tool (Tukker, 2000). Life cycle assessment (LCA) is a specific elaboration of an environmental evaluation framework which differs fundamentally from the above techniques. It is a systematic tool for the identifications, quantifications and evaluations of a product, process or activity, by measuring full range of environmental impacts related to the input and output flows, energy and materials exhaustions and wastes disposal (Guinée, 2002). LCA has been regulated by the ISO 14040 and 14044 since 2006. LCA methodology has been recently applied to quantify and assess environmental impacts of polymer materials (Shen and Patel, 2008), nanomaterials (Pourzahedi and Eckelman, 2015), building materials (Silvestre et al., 2014) and the process industry (Jacquemin et al., 2012), and leather manufacturing. In the newly reported chrome-free tanning process, carbon footprinting calculated by global warming potential (GWP), toxicity indicators and energy consumption have been used for the environmental assessment from LCA perspective; nevertheless, physical performances along with hazardous substances in the leather are not involved (Xu et al., 2015).

Recently, we have developed a novel wet-white tannage based on the vegetable tannin in combination with Laponite nanoclay (Shi et al., 2016). The key is to create synergistic effects between inorganic nanoclay and organic tannins by solving the diffusion of tanning substances and their bindings with collagen fibers of hides and skins. In the present work, we further optimize the combination wet-white tanning process by comparative studying the synergistic tanning effect of two typical vegetable tannins with Laponite. We focus on the binding stability of Laponite in crust leather during post-tanning procedures and its influence on the physical performances of the wet-white leathers. The environmental impacts of the novel wet-white tanning have been quantified by LCA method for the first time. Our aim is to obtain a comprehensive evaluation for a new tanning process and the leather product which would enable careful tanning material design to rationalize processes and overcome ecological problems.

2. Materials and methods

2.1. Materials

1000 kg pickled cattle hide were taken as raw materials for leather manufacture. The chemicals used for leather processing were all of commercial grade. The dosages of the chemicals were all based on the weight of pickled pelts. Two typical vegetable tannins, tannic acid (TA, tannins content: 81 wt%) and mimosa extracts (tannins content: 72.5 wt%) were purchased from Institute of Chemical Industry of Nanjing Forest Products (China) and Seta SUN (Brazil), respectively. Laponite nanoclay, a synthetic disc-shaped nano-silicate consisting of nanoparticles (diameter: 25–30 nm; thickness: ~1 nm) incorporating an organic pyrophosphate peptiser, was purchased from BYK Additives & Instruments (U.K.).

2.2. Leather processing

Table 1 gives the optimized combination tanning process recipe for wet-white leather according to our previously reported methods (Shi et al., 2016). Specifically, in the de-pickling procedure, the pH values of pelts are adjusted to 4.5. And 15% vegetable tannin and 3% Laponite are added for 4 h and 3 h tanning, respectively, with the final pH of the tanning float adjusted at 3.5. Then the wet-white leather are piled overnight. The tanning recipe for chrome tanned leather is also listed in Table 2 as the control. The pickled pelts are tanned with 8% chrome powder for 2 h. Then the float are basified to pH 3.8–4.2 for another 3 h to complete the tanning for wet-blue.

Shrinkage temperature (T_s), a measure of hydrothermal stability of the crust leather (Covington, 1997), was determined by a shrinkage tester using the official method (ASTM, 2004). A 10 mm × 60 mm specimen was cut out from the leather sample and was inserted into the bath of water. Then the bath was heated at a rate such that the rise in temperature was kept at 2 °C per minute. The temperature at the first definite sign of shrinking was recorded.

Table 1
Process recipe for wet-white leather.

Procedures	Chemicals	Amount (%)	Time (min)	Remarks
De-pickling	H$_2$O	100		25 °C
	NaHCO$_3$	1–2	60	Check pH 4.5
Tanning	Vegetable tannin	15	240	
	Nanoclay	4	180	
	HCOOH	0.3–0.5	60	Check pH 3.5 and $T_s \geq 85$ °C, piled overnight

Table 2
Process recipe for conventional chrome tanned leather.

Procedures	Chemicals	Amount (%)	Time (min)	Remarks
Pickling	H$_2$O	100		25 °C
	NaCl	8	10	Check pH 2.8–3.0
	H$_2$SO$_4$	1.25	90	
Tanning	Chrome powder	8	120	Check the penetration
	H$_2$O	100		40 °C
	NaHCO$_3$	1.5	3 × 20 + 120	$T_s \geq 95$ °C, piled overnight

The experimental plot was obtained from average of three samples.

2.3. ICP-AES measurements

The content of Laponite in the wet-white crust leather and its residue in tanning floats were monitored with inductively coupled plasma-atomic emission spectrometry (ICP-AES) technique by measuring the Mg trace as the characteristic element of Laponite (Metreveli et al., 2005). 0.1 g crust leather samples and 2.0 mL floats taken from different wet-white tanning procedures were digested by 10 mL HCl/HNO$_3$ solution (v/v, 3:1) for 2 h at 110 °C, respectively. The digested solution was filtered and the filtrate was then diluted with deionized water. 5 mL analytical solution was used for the measurements of Mg^{2+} concentration by an ICP-AES spectrometer (Optima 2100DV, Perkin-Elmer, USA) equipped with an auto sampler. The analyte wavelength of Mg was 285.2 nm. The adsorption capacity and the residue of Laponite were calculated according to its composition and expressed as mg/g of the crust and mg/L of the float, respectively. The experimental plot was obtained from average of three samples.

2.4. SEM-EDX analysis

The samples were taken from the grain, middle and flesh layers of the wet-white crust leather, respectively. After lyophilized at −43 °C in a freeze dryer (Alpha 1-2 LD, Christ, Germany) for 24 h, the samples were cut into the specimens with a thickness of 1.0 mm by a microtome (CM1900, Leica, Germany) before the observations. The morphologies of leather specimens were recorded by a scanning electron microscope (JSM-7500F, JEOL, Japan) at an accelerating voltage of 15 kV. The relative elemental compositions of the specimens were confirmed by a coupled energy dispersive X-ray spectroscopy (EDX) detector.

2.5. Physical performance and hazardous substances measurements

Physical performance measurements of wet-white crust and finished leather were carried out with the approaches recommended by the International Union of Leather Technologists and Chemists Societies (IULTCS). Tensile strength and percentage extension, tear load and resistance to grain cracking of the crust and finished leather were tested using IULTCS/IUP 6 (in accordance with ISO 3376: 2011), IULTCS/IUP 8 (in accordance with ISO 3377-2: 2016) and IULTCS/IUP 12 (in accordance with ISO 3378: 2002), respectively. The standardized chemical methods recommended by the IULTCS were applied for determining hazardous substances in the wet-white crust and finished leather. Moreover, the contents of limited hazardous substances, Cr(VI) and formaldehyde, were measured with IULTCS/IUC 18 (in accordance with ISO 17075: 2007) and IULTCS/IUC 19 (in accordance with ISO 17226-1: 2008), respectively. The experimental plot was obtained from average of three samples.

2.6. Environmental impacts assessment

For assessing the environmental impacts of the wet-white tanning process, all inventory and procedures data consisting of the inputs and outputs in the tanning procedures were collected and reviewed with mass balances as well as conventional chrome tanning process. A simplified life cycle assessment (LCA) was applied to quantitatively evaluate the environmental impacts of both tanning processes using the standard ISO 14044 (2006).

System boundaries of the comparative life cycle study are shown in Fig. 1. Life cycle inventory (LCI) of the two tanning processes is detailed considering their chemicals and energy consumptions. The inputs and outputs data are depicted in Fig. 2 and Fig. 3, respectively. The study was modeled using GaBi 6 software and the impact categories were evaluated following the CML 2001 methodology (updated in 2013) (Guinée, 2002).

The model takes into account the electricity and thermal energy production, according to China shared-percentage among different used primary energy sources. Chinese electricity mix and thermal energy mix are not available in GaBi6 database; therefore they were developed from Chinese data for tanning regions. Thus, thermal energy was produced from coal (72.9%), gasoline (8.2%), diesel oil

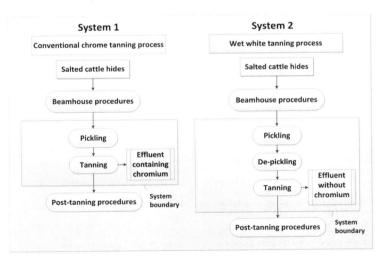

Fig. 1. System boundaries of the comparative life cycle study of conventional chrome tanning process (System 1) and wet-white tanning process (System 2).

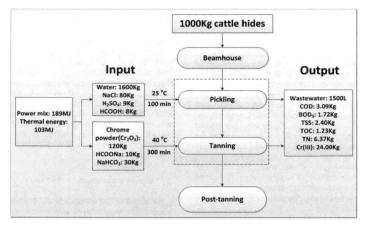

Fig. 2. Flow sheet of chemicals and energy inventory for conventional chrome tanning process.

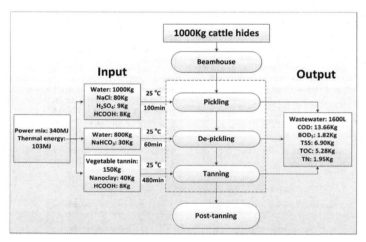

Fig. 3. Flow sheet of chemicals and energy inventory for wet-white tanning process.

(13.3%) and fuel oil (5.7%) and electricity from hydroelectric (17.3%), coal (79.85%), nuclear (1.77%) and wind power plants (1.07%). And, coal is mainly lignite (79.6%) and anthracite (20.4%). Production of chemicals was also considered inside the system boundaries, but due to the lack of some chemicals in GaBi 6 database, the following proxies have been used: acetic acid instead of formic acid and sodium carbonate instead of sodium bicarbonate.

In the case of tannic acid (TA) used in our study, due to lack of data, a similar tannin obtained from patchouli dried leaves in India was considered, which uses ethyl acetate and water as extraction agents. A recycling and re-use of the solvent was also considered in the model. Moreover, Laponite nanoclay was modeled according to its composition in different inorganic components: SiO_2 (54.5 wt%), MgO (26.0 wt%), Li_2O (0.8 wt%), Na_2O (5.6 wt%) and P_2O_5 (4.4 wt%). Waste water treatment (WWT) was modeled considering energy recovery from organic sludge. Data on the WWT plant was taken from the previous reports (Chen et al., 2010; Ban and Zhang, 2011). The process begins with the precipitation of chromium after the addition of sodium hydroxide (if wastewater contains chromium), and then a general biologic waste water treatment including pre-treatment with the addition of ferrous sulfate and polyacrylamide in an aeration tank, secondary treatment and tertiary treatment, with microorganisms to reduce chemical oxygen demand (COD) and other pollution parameters.

3. Results and discussions

3.1. Optimizations of wet-white tannage

Fig. 4 comparatively shows the effect of types of vegetable tannins on the shrinkage temperature (T_s) of wet-white crust leathers. After solo tannic acid (TA) or mimosa tanning with the

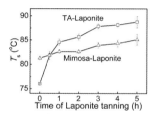

Fig. 4. T_s dependence of wet-white crust leathers on the Laponite tanning time in 15 wt% vegetable tannin combination tannage with 4 wt% Laponite.

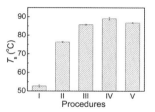

Fig. 6. T_s of the crusts in wet-white leather processing: (I) de-pickling, (II) TA tanning, (III) TA-Laponite combination tanning, (IV) retanning and (V) fatiliquoring.

dosage of 15 wt%, the T_s of the crust leather tanned by mimosa are obviously higher than TA. This is due to the fact that the astringency of condensed vegetable tannin like mimosa is stronger than that of hydrolysable tannin, TA (Covington, 1997). However, further addition of 4 wt% Laponite nanoclay, the crust leathers tanned by the TA-Laponite combination show much higher T_s than those by mimosa-Laponite, as the tanning time prolongs. It is owing to the fact that TA has lower molecule weight (M_w) than mimosa, which can make it more possible to sufficiently penetrate into the leather matrix (Liao et al., 2003). The introduction of Laponite can facilitate the fixation of TA in the collagen fibers by the formation of hydrogen crosslinking between the phenolic hydroxyl groups of TA molecule and the active silanol groups on the surface site of Laponite nano-disk (Shi et al., 2016); whereas, the higher M_w and larger size of mimosa may block the diffusion of Laponite into collagen fibers. Therefore, the synergistic tanning effect of TA-Laponite is much more significant than mimosa-Laponite. Fig. 4 also indicates minor increase in T_s when the time of Laponite tanning is above 3 h. Based on tanning effects and energy consumptions, the TA-Laponite combination tanning for 3 h is thus used for the following assessment.

Fig. 5 shows that Laponite in the grain, middle and flesh layers are all ~32.0 mg/g, indicating its even penetration and binding in the wet-white crust leather. This can be owing to uniform dimensions of Laponite with a thickness of about 1 nm and a diameter of 25–30 nm (Nicolai and Cocard, 2000) and good nano-dispersibility (Rao, 2007), which make it sufficient diffusion from outer float into collagen fiber networks (Gunter, 2007a).

Fig. 6 presents T_s of the crusts in wet-white leather processing. Note that solo 15 wt% TA tanning improves the T_s from 52.8 ± 0.7 °C for the de-pickled pelt to 76.4 ± 0.4 °C. The obvious increase can be owing to hydrogen and hydrophobic bond associations between TA and collagen (Shi et al., 1994). As discussed above, after adding Laponite, T_s of the crusts rise above 86 °C because of the synergistic effects. In the subsequent retanning and fatiliquoring procedures, the resultant crusts exhibit T_s of 89.1 ± 0.7 °C and 87.6 ± 0.3 °C,

respectively, indicating the stability of TA-Laponite combination tanning. Clearly, the T_s of the combination tanned crust leather can meet the requirements for next finishing procedures.

3.2. ICP analysis

Fig. 7 gives the content of Laponite in the crusts (Fig. 7a) and floats (Fig. 7b), respectively, during the wet-white tanning process as measured by ICP technique. The binding capacity of Laponite in the tanned crust leather (Procedure II) is 32.45 ± 1.60 mg/g, i.e., ~80% of Laponite is fixed in the collagen fibers. Its corresponding residual content is 1.73 ± 0.05 mg/L in the tanning float. After retanning and fatiliquoring, some Laponite was washed out (0.5–0.6 mg/L in the floats). This is reasonable because the fixation of tanning materials is known to be reversible under sustained exposure to water (Gunter, 2007b). However, more than 77% is still bound in the crusts to sufficiently maintain the hydrothermal stability.

3.3. SEM-EDX analysis

Fig. 8 shows the microstructures of wet-white crust leather observed by scanning electron microscopy (SEM). Fig. 8a shows the grain layer of the leather, in which the hair pores are clearly visible

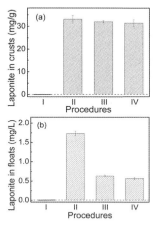

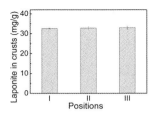

Fig. 5. The content of Laponite nanoclay in the TA-Laponite combination tanned crust leather: (I) grain layer, (II) middle layer and (III) flesh layer.

Fig. 7. The content of Laponite in the wet-white crust leather (a) and its residue (b) in the tanning float during the leather processing: (I) TA tanning, (II) TA-Laponite combination tanning, (III) retanning and (IV) fatiliquoring.

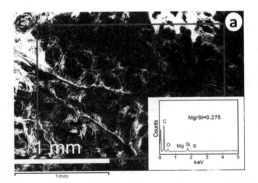
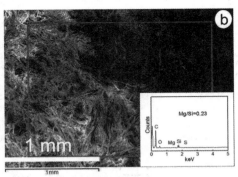

Fig. 8. SEM microphotographs and the corresponding EDX spectrums (the insets) of TA-Laponite combination tanned crust leather: (a) grain layer, (b) middle layer and (c) flesh layer.

Table 4
The content of hazardous substances in the wet-white crust leather and finished leather.

Parameters	Crust	Finished leather	Limitations[a]
Cr(VI) (mg/kg)	Not detected	Not detected	3 mg/kg
Formaldehyde (mg/kg)	Not detected	Not detected	20 mg/kg

[a] The limitations of the detection.

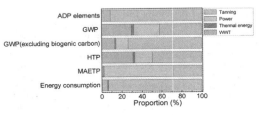

Fig. 9. Contribution of different life cycle stages of the wet-white tanning process to each impact category: abiotic depletion potential (ADP), global warming potential (GWP), GWP excluding biogenic carbon, human toxicity potential (HTP), marine aquatic ecotoxicity potential (MAETP) and energy consumption.

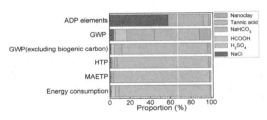

Fig. 10. Contribution of different chemicals used in the wet-white tanning process to each impact category: abiotic depletion potential (ADP), global warming potential (GWP), GWP excluding biogenic carbon, human toxicity potential (HTP), marine aquatic ecotoxicity potential (MAETP) and energy consumption.

without surface depositions of Laponite. The leather specimens in the middle and flesh layers also show similar morphologies with an interwoven structure (Fig. 8b, c). Moreover, the elemental compositions in the corresponding positions of the leather matrix (pink areas in SEM images) have been analyzed by energy dispersive X-ray (EDX) spectroscopy as shown in the insets. In the EDX spectra,

Table 3
Physical properties of the wet-white crust leather and finished leather.

Physical properties	Crust	Finished leather	Norms
Tensile strength (MPa)	36.5 ± 4.2	43.6 ± 2.5	–
Tear strength (N)	81.9 ± 3.0	114.9 ± 11.5	≥40
Elongation at 10 N (%)	31.1 ± 7.5	32.3 ± 7.2	–
Elongation at break (%)	48.5 ± 6.5	55.5 ± 7.3	35–60
Flexing crack resistance	No cracking	No cracking	20,000 flexes with no cracks

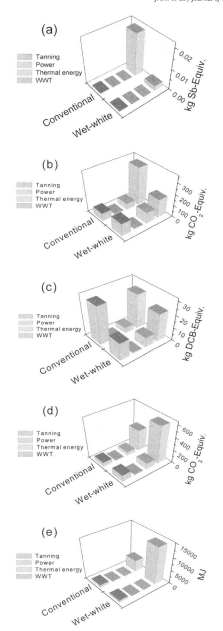

Fig. 11. Comparison between wet-white and conventional chrome tanning process: (a) abiotic depletion potential (ADP), (b) global warming potential (GWP), (c) human toxicity potential (HTP), (d) GWP excluding biogenic carbon and (e) energy consumption.

magnesium (Mg) and silicon (Si), the feature elements of Laponite, show similar contents in the grain, middle and flesh layers, further demonstrating an even distribution of Laponite within the leather matrix. It can be expected that a reasonable good leather properties can be resulted.

3.4. Physical performance measurements

Table 3 presents physical properties of the wet-white crust leather and finished leather tested with the official methods recommended by IULTCS. It can be found that all of these physical properties can meet the Chinese standard requirements for furniture leather (GB/T 16799-2008), indicating that the wet-white tannage can be applied in furniture leather manufacture.

3.5. Hazardous substances results

Table 4 lists the content of limited hazardous substances in the wet-white crust leather and finished leather. Both Cr(VI) and formaldehyde are not detected, showing that the leather products can meet the demands of the EU eco-label for footwear (2009/563/EC).

3.6. Environmental impacts assessment

From the perspective of life cycle assessment (LCA), environmental impact categories of the wet-white tanning process were assessed through the CML 2001 methodology (revised on 2013). Fig. 9 gives the contribution of four factors to the impact categories. The results show that tanning is the main contributor to all impact categories except for marine aquatic ecotoxicity potential (MAETP) in which power (electricity production) is more significant. As to the substances in tanning process (Fig. 10), the most significant contributor for all impact categories is tanning agent, i.e. tannic acid, except for abiotic depletion potential (ADP).

Fig. 11 comparatively shows the differences of environmental impact categories between wet-white and conventional chrome tanning process. As can be seen in Fig. 11a, the wet-white tanning process has much lower impact in ADP with a reduction of 88.2%, mainly due to the fact that chromium derivatives have a problem of abiotic resource depletion. Also, the wet-white tanning process performs better in global warming potential (GWP) impact (Fig. 11b) and the impact decreases from 463.2 kg CO_2 equivalent to 386.0 kg CO_2 equivalent by 16.7% than the conventional one. Moreover, the human toxicity potential (HTP) impact results (Fig. 11c) show a great preference for the novel wet-white tanning process by reducing a 28.2% compared to the conventional one. This reduction is mainly related to the decrease in the impact from WWT followed by the decrease in the tanning contributor owing to the absence of chromium, since TA tannin can be easily biodegraded with low toxicity and Laponite nanoclay does not show acute toxicity due to its nature of silicate (Field and Lettinga, 1992; Gaharwar et al., 2011).

However, other environmental impacts of the wet-white tanning process are not better than the conventional chrome tanning process. The GWP excluding biogenic carbon (Fig. 11d) results also show that tanning, energy consumption (power) and wastewater treatment (WWT) are higher in the wet-white tanning process due to the use of organic tanning materials. And energy consumption (Fig. 11e) is also much higher due to more tanning time requirements in the combination tanning. The results imply that the use of environmental friendly vegetable tannins is not always harmless to environment. Reducing the use of vegetable tanning material (or partially replace it with highly reactive mineral tanning agents), as well as decreasing electricity requirements, may mitigate these disadvantages. The LCA assessments on three typical tanning processes based on chrome, aldehyde and vegetable tannin have been previously reported by Ecobilan S.A. (BLC Leather Technology Centre Report 002). It was said that there is no fundamental difference in environmental impacts between these tanning technologies, and each tanning technology has strengths and weakness (BLC, 2013). Indeed, our LCA results reveal that the wet-white tanning process does not show absolute advantages in all the environmental impacts.

4. Conclusions

A novel wet-white leather manufacture based on vegetable tannin and Laponite nanoclay combination tannage has been established in this work. The 15 wt% tannic acid combination with 4 wt% Laponite tanning for 3 h can give the leather with T_s above 86 °C. Laponite can penetrate evenly into leather matrix and bind tightly with collagen fibers. Moreover, the resultant leathers can meet the official standard requirements of physical properties for furniture leather, with no limited hazardous substances, Cr(VI) and free formaldehyde detected. LCA results indicate that tanning process is the main contributor to environmental impact categories. And the wet-white tanning process exhibits reduced environmental impacts in ADP, GWP and HTP impacts, but higher GWP impact excluding biogenic carbon and energy consumption than conventional chrome tanning. A reduction of both, vegetable tannin and electricity, consumptions would mitigate these disadvantages. The comprehensive evaluation on the leather properties and the environmental impacts from a novel tanning process is realistic and significant for the exploration of new tanning materials and the related process design.

Acknowledgements

The financial support of National Natural Science Foundation (NNSF) of China (21176159, 21476148), and National High-tech Research and Development Program (863 program) of China (2013AA06A306) is gratefully acknowledged.

References

ASTM, 2004. Annual Book of ASTM Standards, 15-04. American Society for Testing & Materials, pp. 638–640.
Ban, F.C., Zhang, T., 2011. Design example of wastewater treatment plant of leather industry base. Technol. Water Treat. 37, 124–126.
Bridges, J., 2003. Human health and environmental risk assessment: the need for a more harmonised and integrated approach. Chemosphere 52, 1347–1351.
British Leather Center (BLC), 2013. Environmental Impact of Different Tanning Technologies. http://www.blcleathertech.com/blog/environmental-impact-of-different-tanning-technologies/2013/04/24/.
Brunner, P.H., Rechberger, H., 2004. Practical handbook of material flow analysis. Int. J. Life Cycle Assess. 9, 337–338.
Burgess, A.A., Brennan, D.J., 2001. Application of life cycle assessment to chemical processes. Chem. Eng. Sci. 56, 2589–2604.
Chandra Babu, N.K., Asma, K., Raghupathi, A., Venba, R., Ramesh, R., Sadulla, S., 2005. Screening of leather auxiliaries for their role in toxic hexavalent chromium formation in leather-posing potential health hazards to the users. J. Clean. Prod. 13, 1189–1195.
Chen, Y.P., Tan, J.L., Tang, H.L., Lu, T.L., Yang, D.Q., 2010. Technical design and application analysis of tannery wastewater treatment. Biomass Chem. Eng. 44, 32–35.
Covington, A.D., 1997. Modern tanning chemistry. Chem. Soc. Rev. 26, 111–126.
Covington, A.D., 2009. Tanning Chemistry: The Science of Leather. Royal Society of Chemistry, pp. 204–205.
Crudu, M., Deselnicu, V., Deselnicu, D.C., Albu, L., 2014. Valorization of titanium metal wastes as tanning agent used in leather industry. Waste Manag. 34, 1806–1814.
Erol, P., Thöming, J., 2005. ECO-design of reuse and recycling networks by multi-objective optimization. J. Clean. Prod. 13, 1492–1503.
Field, J., Lettinga, G., 1992. Toxicity of tannic compounds to microorganisms. In: Plant Polyphenols. Springer, pp. 673–692.
Gaharwar, A.K., Schexnailder, P.J., Kline, B.P., Schmidt, G., 2011. Assessment of using Laponite® cross-linked poly (ethylene oxide) for controlled cell adhesion and mineralization. Acta Biomater. 7, 568–577.
Guinée, J.B., 2002. Handbook on life cycle assessment operational guide to the ISO standards. Int. J. Life Cycle Assess. 7, 311–313.
Gunter, R., 2007a. From Collagen to Leather - the Theoretical Background. BASF Service Center Media and Communications, Ludwigshafen, pp. 42–45.
Gunter, R., 2007b. From Collagen to Leather - the Theoretical Background. BASF Service Center Media and Communications, Ludwigshafen, p. 47.
Hansen, M., Johansen, J., Menne, T., 2003. Chromium allergy: significance of both Cr(III) and Cr(VI). Contact Dermat. 49, 206–212.
Jacquemin, L., Pontalier, P.Y., Sablayrolles, C., 2012. Life cycle assessment (LCA) applied to the process industry: a review. Int. J. Life Cycle Assess. 17, 1028–1041.
Jay, S., Jones, C., Slinn, P., Wood, C., 2007. Environmental impact assessment: retrospect and prospect. Environ. Impact Assess. Rev. 27, 287–300.
Krishnamoorthy, G., Sadulla, S., Sehgal, P.K., Mandal, A.B., 2013. Greener approach to leather tanning process: D-Lysine aldehyde as novel tanning agent for chrome-free tanning. J. Clean. Prod. 42, 277–286.
Liao, X.P., Lu, Z.B., Shi, B., 2003. Selective adsorption of vegetable tannins onto collagen fibers. Ind. Eng. Chem. Res. 42, 3397–3402.
Liu, Y.S., Chen, Y., Yao, J., Fan, H.J., Shi, B., Peng, B.Y., 2010. An environmentally friendly leather-making process based on silica chemistry. J. Am. Leather Chem. Assoc. 105, 84–93.
Metreveli, G., Kaulisch, E.M., Frimmel, F.H., 2005. Coupling of a column system with ICP-MS for the characterisation of colloid-mediated metal (loid) transport in porous media. Acta Hydrochim. Hydrobiol. 33, 337–345.
Nicolai, T., Cocard, S., 2000. Light scattering study of the dispersion of laponite. Langmuir 16, 8189–8193.
Pourzahedi, L., Eckelman, M.J., 2015. Environmental life cycle assessment of nanosilver-enabled bandages. Environ. Sci. Technol. 49, 361–368.
Rao, Y., 2007. Gelatin-clay nanocomposites of improved properties. Polymer 48, 5369–5375.
Shen, L., Patel, M.K., 2008. Life cycle assessment of polysaccharide materials: a review. J. Polym. Environ. 16, 154–167.
Shi, B., He, X., Haslam, E., 1994. Gelatin-polyphenol interaction. J. Am. Leather Chem. Assoc. 89, 96–102.
Shi, J.B., Ren, K.S., Wang, C.H., Wang, J., Lin, W., 2016. A novel approach for wet-white leather manufacture based on tannic acid-Laponite nanoclay combination tannage. J. Soc. Leather Technol. Chem. 100, 25–30.
Shi, J.B., Zhou, Y.L., Li, X.P., Lin, W., 2013. A novel combination tanning based on tannic acid and attapulgite nanoclay. China Leather 3, 1–5.
Silvestre, J.D., de Brito, J., Pinheiro, M.D., 2014. Environmental impacts and benefits of the end-of-life of building materials - calculation rules, results and contribution to a "cradle to cradle" life cycle. J. Clean. Prod. 66, 37–45.
Tukker, A., 2000. Life cycle assessment as a tool in environmental impact assessment. Environ. Impact Assess. Rev. 20, 435–456.
Wang, Y.N., Zeng, Y.H., Zhou, J.F., Zhang, W.H., Liao, X.P., Shi, B., 2016. An integrated cleaner beamhouse process for minimization of nitrogen pollution in leather manufacture. J. Clean. Prod. 112, 2–8.
Xu, X.Y., Baquero, G., Puig, R., Shi, J.B., Sang, J., Lin, W., 2015. Carbon footprint and toxicity indicators of alternative chromium-free tanning in China. J. Am. Leather Chem. Assoc. 110, 130–137.
Zhang, C.X., Lin, J., Jia, X.J., Peng, B.Y., 2016. A salt-free and chromium discharge minimizing tanning technology: the novel cleaner integrated chrome tanning process. J. Clean. Prod. 112, 1055–1063.

A Novel Approach for Wet-White Leather Manufacture Based on Tannic Acid-Laponite Nanoclay Combination Tannage

SHI JIABO[1], REN KESHUAI[2], WANG CHUNHUA[1], WANG JIE[2] and LIN WEI[1]*

[1] *National Engineering Laboratory for Clean Technology of Leather Manufacture, Sichuan University, Chengdu, Sichuan, China*

[2] *Department of Biomass and Leather Engineering, Key Laboratory of Leather Chemistry and Engineering of Ministry of Education, Sichuan University, Chengdu, Sichuan, China*

Abstract

A novel wet-white tanning approach based on combination tanning with tannic acid (TA) and Laponite nanoclay has been proposed. Thermal stability measurements show that the introduction of Laponite nanoclay gives an obvious increase in the shrinkage temperature (T_s) of TA-tanned leather, implying the presence of the synergistic effect between TA and Laponite nanoclay in the combination tanning process. The optimized tanning system, *i.e.* 20% TA combination tanned with 3% Laponite in one-bath for 3 hours at 25°C and final pH ~3.5, confers wet-white leathers with T_s above 89°C, and good storage stability. The introduction of Laponite nanoclay facilitates the fixation of TA in the wet-white leather. Scanning electron microscopic (SEM) results show that the combination tanned leathers exhibit fine isolated collagen fibre networks in comparison with that of solo TA and solo Laponite-tanned leather. The presence of Laponite nanoclay not only improves the mechanical strength of the resulting leather, but also gives an ultraviolet protection property.

摘要

研究并建立了一种新型的基于单宁酸-Laponite结合鞣的白湿革鞣法。湿热稳定性测试结果表明，在单宁酸-Laponite结合鞣中单宁酸与Laponite存在着协同鞣制作用。当单宁酸用量和Laponite用量分别为皮重20%和3%时，采用同浴结合鞣法鞣制浸酸山羊皮，在终点pH为~3.5、鞣制时间为3h条件下，结合鞣白湿革收缩温度可达89°C以上，耐储存性好。扫描电镜分析(SEM)结果表明，相比单宁酸单独鞣革，结合鞣白湿革纤维具有较松散的微观结构。Laponite的加入不仅可以明显提高坯革对单宁酸的吸收效果，而且赋予结合鞣坯革良好的物理机械性能，同时赋予结合鞣坯革抗紫外性。

1. INTRODUCTION

Currently, there is a globally growing requirement for wet-white leather products, including automotive leathers, upholstery leather and garment leather with considerable high-quality and high value-added characters.[1,2] The wet-white tanning approach produces non-polluting and non-toxic solid wastes in the splitting and shaving process without the possible hazard of Cr(VI) formation. This reveals the by-products of leather manufacture to be potentially valuable.[3] Furthermore, wet-white tanning presents substantial environmental benefits because of its significant reductions in hazardous solid waste disposal and the level of chemicals in the effluents. Therefore, wet-white tanning is considered as a feasible alternative to conventional chrome tanning for chrome free leather manufacture.[2,4]

A great variety of wet-white leather products can be produced by non-chrome tanning agents. More recently, hydrolysable vegetable tannins are deemed as a potential and feasible wet-white tanning agents.[5] Growing attention has been drawn to tannic acid (TA) because of its suitable molecular weight and sufficient functional groups.[6] It can adequately be absorbed by collagen fibre due to the cross-linking effects primarily through the formation of multiple hydrogen bonds.[7,8] Moreover, vegetable tannins are an environmentally-friendly and renewable natural material.[9] However, there are still several drawbacks in solo TA tannage, such as producing the leathers with limited hydrothermal stability and excessive fullness as well as lack of necessary softness which are attributed to the characteristics of vegetable tannage.[5] Wet-white tannage based on TA combination with other tanning materials, especially silica and silicate have been widely investigated to improve these drawbacks.[3,10] In our previous study, wet-white tannage using TA in combination with attapulgite nanoclay was conducted and the results show that TA synergized with attapulgite nanoclay can accelerate the uptake of TA in the leathers and improve the hydrothermal stability of leather.[11] However, poor dispersion of the nanoclay limits its penetration into the leather's matrix and affects the performances of leathers. Therefore, nanoclays with appropriate size, good dispersibility and sufficient reactivity could provide approaches to solve these problems.

* Corresponding author: E-mail: wlin@scu.edu.cn

Laponite nanoclay is a synthetic nano-layered silicate which has a relatively uniform disc-like shape with the diameter of 25nm and the thickness of 1nm.[12] Owing to its non-toxicity, security to human health and novel colloidal properties, Laponite nanoclay has been widely utilized in many formulations of industrial waterborne products.[13] It is reported that the association of gelatin and Laponite nanoclay is facilitated by a strongly attractive interaction potential which can lead to preferential binding of the biopolymer chains to negatively charged face of the nanoclay.[14] In addition, Laponite is well exfoliated, dispersed, and parallel to the film plane in gelatin-Laponite nanocomposite films and can also give an enhanced mechanical property to the film.[15] These studies inspired us to explore novel wet-white tannage with Laponite nanoclay as a tanning agent.

In this work, we have investigated the introduction of the Laponite nanoclay into tannic acid (TA) based wet-white tanning. The parameters of the combination tannage have been optimized. Physical and strength properties of the leathers have been characterized. Our aim is to develop novel and eco-friendly tanning materials and the corresponding feasible tannage for wet-white leather processing.

2. EXPERIMENTAL PROCEDURE

2.1 Materials

Pickled goatskins with an average area of 4-5 square metres were taken as raw materials for wet-white leather manufacture. The dosages of the chemicals used for leather processing were all based on the weight of pickled pelts. Hydrolysable vegetable tannin, tannic acid (TA) was supplied from the Institute of Chemical Industry of Nanjing Forest Products (China). Laponite RDS was purchased from Rockwood Additives Ltd. Co. (U.K.). The chemicals used for leather processing were of commercial grade and the others were of analytical grade.

2.2 Solo TA or Laponite tanning process

Pickled pelts were used for solo TA tanning trials according to the reported methods.[16] The pelts were first immersed in 100% pickle liquors at 25°C and de-pickled by a proper amount of $NaHCO_3$ solution (1.5%, w/w) for 1 hour to adjust the pH of pelts to ~4.5. The de-pickled pelts were tanned with 5%-30% TA, and the drum was run for 4 hours to give complete penetration of TA. The final pH was adjusted to pH3.0 using formic acid solution (1.0%, w/w). The crust leather was then washed with 200% water and piled overnight. Shrinkage temperature (T_s) of the leather was tested and recorded by a shrinkage tester. Besides, for the solo Laponite nanoclay tanning, pickled pelts were tanned with 1% ~ 5% Laponite at 25°C for 3 hours, respectively. The obtained leathers were washed with 200% water and piled overnight. T_s of the leathers were measured.

2.3 TA-Laponite combination tanning process

Pickled pelts were immersed in 100% pickle liquor at 25°C and de-pickled according to the approach in Section 2.2. The de-pickled pelts were first tanned with TA at levels ranging from 10% to 30%, run 4 hours for complete penetration. The pH of the float was ~3.0. The crust leathers were then tanned with 1% to 5% Laponite nanoclay. The drum was run for another 1-5 hours and the final pH of leathers was adjusted to ~3.5 by formic acid solution (1.0%, w/w). The obtained leathers were then washed with 200% water and piled overnight. T_s of the leathers were then measured. The spent tan liquors were collected to determine the uptake of TA in the leathers. The post-tanning processing was conducted according to normal procedures including neutralization, retanning and fatliquoring.[10] The resultant wet-white leathers were tested for physical strength properties and anti-UV property measurements.

2.4 Determination of shrinkage temperature

Shrinkage temperature (T_s) was measured by a shrinkage tester by the ASTM method.[17] A 10mm x 60mm specimen was cut out from each leather sample and was inserted into the bath of water. Then the bath was heated at a rate such that the rise in temperature was kept at 2°C per minute. The temperature at the first definite sign of shrinking was recorded.

2.5 Determination of the uptake of TA

10mL spent tan liquors were collected and centrifuged at 5000rpm for 30 minutes and the supernatants were collected for total organic carbon (TOC) analysis. 1mL sample was diluted to 100mL with deionized water. TOC measurements were performed by a high temperature TOC/TNb analyzer (LiquiTOC/TNb, Elementar, Germany) coupled with automatic sampling instrument.[18] The uptake of TA was expressed as the TOC removal and calculated as the following formula:

$$\text{Uptake of TA} = (C_0 - C_t) / C_0 \times 100 \quad (1)$$

where C_0 and C_t are the concentration of TA expressed as TOC value in the spent tan liquor before and after the tanning process, respectively.

2.6 Scanning electron microscopic analysis of crust leathers

Leather specimens were firstly frozen at -50°C and lyophilized in a freeze dryer (Alpha 1-2 LD, Christ, Germany) for 24 hours. The specimens were embedded in paraffin wax, cut into specimens and immersed in 100% xylene for 24 hours to remove the paraffin wax. Excess xylene was removed by rinsing in 100% ethanol. Then, the specimens were dried for 12 hours at room temperature and inserted into a desktop scanning electron microscope (Phenom Pro, Phenom-World Company, China). SEM microphotographs for the grain surface and the cross section of leathers were obtained by operating the microscope at an accelerating voltage of 5kV.

2.7 Physical properties examination

Leather specimens for physical properties testing were sampled according to IULTCS methods.[19] The specimens were first conditioned at ambient temperature with relative humidity of 65 ± 2% for 48 hours. Then physical properties of the leathers, such as tensile strength, tear strength and elongation at break were tested by the standard methods.[20,21] After the leathers were stored for 6 months (180d), these properties were tested again by the same methods.

2.8 Anti-UV property measurements

Anti-UV property of the leather was characterized using UV-Vis diffuse reflectance measurement by a UV-Vis-NIR spectrophotometer (UV-3600 with an integrating sphere attachment, Shimadzu, Japan) with barium sulfate as a reference. UV-Vis absorption spectra of the leathers were recorded in the UV range from 200nm~400nm.

3. RESULTS AND DISCUSSION

3.1 Optimization of solo TA tannage

Effect of TA dosage on the shrinkage temperature (T_s) and the uptake of TA in solo TA-tanned crust leathers are presented in Figure 1. It is clear that both T_s (Fig. 1a) and the uptake of TA (Fig.1b) increase markedly with increasing dosage of TA. This is understandable because of the hydrogen and hydrophobic bond associations between phenolic hydroxyl groups of TA molecule and functional groups of collagen molecule.[22] Note that the uptake of TA is 84.9% when its dosage is 20% based on the weight of pickled pelts, indicating that TA can be strongly absorbed by collagen fibres.[8] Further increasing TA does not lead to obvious increase in T_s. Therefore, 20% TA was used in the subsequent combination tanning experiments and it only confers, to the crust leather, a T_s of 67°C.

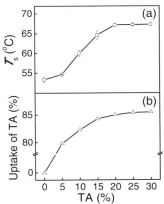

Figure 1. T_s of crust leathers (**a**) and the uptake of TA (**b**) as a function of the dosage of TA in the solo TA tanning.

3.2 Optimization of TA-Laponite nanoclay combination tannage

Figure 2 shows the T_s of TA-Laponite combination-tanned crust leathers. Note that the solo Laponite tanning with 1% to 5% dosage gives an increase in T_s of less than 5°C. It is known that the interactions between collagen molecules and Laponite nanoclay can primarily be related to the dehydration of nano-silicates to collagen fibres and the strong electrostatic effect,[14,23] which seem to have a minor contribution to the improvement of hydrothermal stability of the leather. In contrast, the TA-Laponite nanoclay combination tannage provides a substantial increase in T_s with increasing dosage of Laponite. The change of shrinkage temperature, ΔT_s, is much higher than that of solo TA or Laponite tannage, showing the synergistic tanning effect of TA-Laponite. Furthermore, 20% TA-Laponite combination tanning obtains the highest T_s for the crust leathers, nearly 90°C. The results reveal that the dosage of TA is closely related to the performance of the novel combination tannage. Either low or high concentrations of TA in the tanning float cannot help to enhance the cross-linking between TA, Laponite nanoclay and collagen fibres. The former may cause the insufficient binding or cross-linking, while the latter can lead to flocculation between excessive TA and Laponite nanoclay in the outer float.

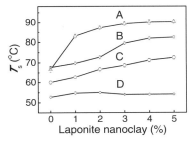

Figure 2. T_s of the leathers produced by TA-Laponite nanoclay combination tannage (**A**: 20% TA-Laponite; **B**: 30% TA-Laponite; **C**: 10% TA-Laponite; **D**: solo Laponite tanning).

Figure 3 further indicates that in the combination tanning with 20% TA and Laponite nanoclay, appropriate nanoclay can facilitate the fixation of TA in the crust leather. When the dosage of Laponite nanoclay is above 3%, the uptake of TA slowly increases; meanwhile, the corresponding T_s has almost no effect (Fig. 2). Therefore, 3% Laponite combined with 20% TA is mostly optimized from the point of view of tanning effects and efficiency, and thus was used for further investigations. The synergistic tanning mechanism of TA molecule and Laponite nanoclay can be due to the multifunctional cross-linking effect of the nanoclay, *i.e.*, Laponite can not only form a linkage with collagen molecule by hydrogen bonding but also can cross-link with phenolic hydroxyl groups of TA molecule.[24,25,26]

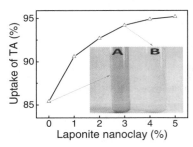

Figure 3. Relation of uptake of TA to the dosage of Laponite nanoclay in the combination tanning. The insert shows the appearance of spent tan liquors (**A**: solo 20% TA; **B**: 20% TA-3% Laponite nanoclay). [The liquor in tube **A** has a moderate yellow-brown colour, that in tube **B** is virtually clear :Ed.]

Figure 4a shows the effect of final pH on the T_s of wet-white leather. After the introduction of Laponite nanoclay, the pH of the tanning float increases from ~3.0 to ~5.0, which is due to the fact that the pH of Laponite aqueous solution itself is nearly pH10. However, the T_s at the final pH5.0 is only 82°C. When the final pH value is decreased to ~3.5, a T_s of around 90°C is achieved, indicating stronger combination tanning effects of TA and Laponite at pH3.0~3.5. The reason may be that it is close to the isoelectronic point of TA, and there exists the physical deposition of TA within collagen fibres.[27] In addition, as given in Figure 4b, the T_s of wet-white leather is greatly enhanced after adding Laponite for 1 hour. After the tanning duration of 3 hours, the hydrothermal stability of the leather is nearly constant. Therefore, the final pH and the Laponite tanning time in the combination tannage are optimized as pH ~3.5 and 3 hours, respectively.

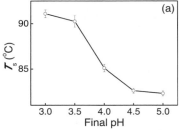 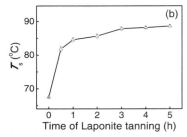

Figure 4. Final pH (**a**) and time of Laponite tanning (**b**) dependence of T_s of wet-white leathers by 20% TA and 3% Laponite nanoclay at 25°C.

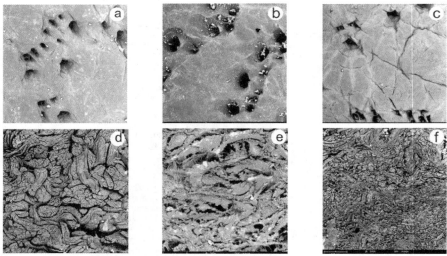

Figure 5. SEM morphologies of grain sides (**a**, **b**, and **c**) and cross sections (**d**, **e**, and **f**): 20% TA-tanned crust leather (**a** and **d**), 3% Laponite-tanned crust leather (**b** and **e**) and the combination-tanned wet-white leather (**c** and **f**).

3.3 Scanning electron microscopic analysis of crust leathers

Figure 5 shows the comparative SEM microphotographs of the leathers by different tannages. The grain structure of all the leathers (Figs. 5a, 5b, and 5c) seems clean without any damage and the hair pores are clearly visible without surface deposition of TA or Laponite, indicating the good uptake of tanning materials. While the cross section structures (Figs. 5d, 5e, and 5f) exhibit a distinct difference in the fibre morphologies. The fibres of TA-Laponite combination-tanned crust leather show fine isolated networks (Fig. 5f). Whereas, solo TA-tanned crust exhibits dense and thick collagen fibre bundles (Fig. 5d), which is primarily attributed to the filling and astringent nature of vegetable tannage.[28] Specifically, solo Laponite-tanned crust leather shows a thick and coarse fibre network owing to the nature of a silicate tannage (Fig. 5e). The results indicate that the synergistic combination tanning of TA and Laponite can overcome the problem of loosening of collagen fibres by solo TA or solo silicate.

3.4 Physical strength properties of leathers

The physical properties of the wet-white leathers including tensile strength, tear strength and elongation at break are shown in Table I. It is evident that the wet-white leather tanned with 20% TA and 3% Laponite combination tannage exhibits improved physical strength properties over those of solo 20% TA. This is in accordance with the hydrothermal stability and the uptake of TA. Moreover, these physical strength properties showed virtually no decrease after storage for 6 months, indicating that the combination-tanned wet-white leather possesses better storage stability.

3.5 Anti-ultraviolet property of leathers

Figure 6 shows the UV transmittances of wet-white leathers and solo TA-tanned leather at the wavelength of 200~400nm. The wet-white leather exhibits higher absorbance or lower transmittance in the near UV region. In other words, it has better anti-UV property owing to the residence Laponite nanoclay to the UV irradiation. The reason is identified as scattering of the incident light by Laponite nanoclay.[29]

TABLE I
Physical strength properties of the wet-white leathers before and after 6 months storage

Properties		Solo TA	TA-Laponite
Tensile strength (MPa)	0d	14.0 ± 3.5	16.1 ± 2.4
	180d	11.8 ± 1.5	15.6 ± 2.7
Tear strength (N/mm)	0d	57.6 ± 3.0	83.1 ± 1.0
	180d	50.3 ± 3.5	81.5 ± 2.8
Elongation at 5N (%)	0d	21.5 ± 2.1	42.1 ± 2.7
	180d	18.5 ± 1.8	38.1 ± 4.5
Elongation at break (%)	0d	50.5 ± 1.6	54.5 ± 5.2
	180d	40.6 ± 3.6	52.3 ± 3.9

4. CONCLUSIONS

A novel combination tanning approach based on TA and Laponite nanoclay for wet-white leather manufacture has been established in this work. The results show that the TA-Laponite combination tannage confers to the crust leather a substantial increase in T_s and the enhanced uptake of TA with increasing dosage of Laponite, exhibiting the synergistic effect of TA and Laponite. The optimized combination tannage is to use 20% TA and 3% Laponite nanoclay in the same float at final pH ~3.5, which gives the leather a T_s above 89°C. SEM analysis reveals that the combination tanned crust leathers have fine isolated collagen fibre network structures. Moreover, Laponite nanoclay improves the strength properties of wet-white leather with good storage stability. The leather also exhibits the enhanced anti-UV property.

ACKNOWLEDGMENTS

The authors greatly appreciate the financial support from National Natural Science Foundation of China (21176159) and National High-tech Research and Development Program of China ('863' project, Project number: 2013AA06A306).

(Received October 2015)

in association with China Leather

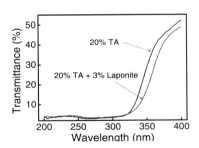

Figure 6. Ultraviolet transmittance spectra of leathers.

References

1. Taylor, M., Lee, J., Bumanlag, L. *et al.*, Treatments to enhance properties of chrome-free (wet-white) leather. *J. Amer. Leather Chem. Ass.*, 2011, **106**(2), 35-43.
2. Wolf, G., Breth, M., Carle, J. *et al.*, New developments in wet-white tanning technology. *J. Amer. Leather Chem. Ass.*, 2001, **96**(4), 111-119.
3. Saravanabhavan, S., Fathima, N. N., Rao, J. R. *et al.*, Combination of white minerals with natural tannins-Chrome-free tannage for garment leathers. *J. Soc. Leather Technol. Chem.*, 2004, **88**(2), 76-81.
4. DasGupta, S., Novel eco-friendly approaches for the production of upholstery leather. *J. Amer. Leather Chem. Ass.*, 2009, **104**(3), 92-102.
5. Musa, A. E. and Gasmelseed, G. A., Combination tanning system for manufacture of shoe upper leathers: cleaner tanning process. *J. Soc. Leather Technol. Chem.*, 2012, **96**(6), 239-245.

6. Olsen, S. N., Bohlin, C., Murphy, L. et al., Effects of non-ionic surfactants on the interactions between cellulases and tannic acid: a model system for cellulase-polyphenol interactions. *Enzyme and Microbial Technology*, 2011, **49**(4), 353-359.
7. Heijmen, F., DuPont, J., Middelkoop, E. et al., Cross-linking of dermal sheep collagen with tannic acid. *Biomaterials*, 1997, **18**(10), 749-754.
8. Liao, X. P., Lu, Z. B., and Shi, B., Selective adsorption of vegetable tannins onto collagen fibres. *Industrial & Engineering Chemistry Research*, 2003, **42**(14), 3397-3402.
9. Castell, J. C., Sorolla, C. S., Jorba, M. et al., Tara (*Caesalpinia spinosa*): the sustainable source of tannins for innovative tanning processes. *J. Amer. Leather Chem. Ass.*, 2011, **108**(6), 221-230.
10. Fathima, N. N., Saravanabhavan, S., Rao, J. R. et al., An eco-benign tanning system using aluminium, tannic acid, and silica combination. *J. Amer. Leather Chem. Ass.*, 2004, **99**(2), 73-81.
11. Shi, J. B., Zhou, Y. L., Li, X. P. et al., A novel combination tanning based on tannic acid and attapulgite nanoclay. *China Leather*, 2013, **42**(3), 1-5.
12. Nicolai, T. and Cocard, S., Light scattering study of the dispersion of laponite. *Langmuir*, 2000, **16**(21), 8189-8193.
13. Aouada, F. A. and Mattoso, L. H. C., A simple procedure for the preparation of laponite and thermoplastic starch nanocomposites: structural, mechanical and thermal characterizations. *Journal of Thermoplastic Composite Materials*, 2013, **26**(1), 109-124.
14. Pawar, N. and Bohidar, H. B., Surface selective binding of nanoclay particles to polyampholyte protein chains. *Journal of Chemical Physics*, 2009, **131**(4), 225-229.
15. Rao, Y. Q., Gelatin-clay nanocomposites of improved properties. *Polymer*, 2007, **48**(18), 5369-5375.
16. Fathima, N. N., Aravindhan, R., Rao, J. R. et al., Tannic acid-phosphonium combination: a versatile chrome-free organic tanning. *J. Amer. Leather Chem. Ass.*, 2006, **101**(5), 161-168.
17. ASTM International. Annual Book of ASTM Standard, West Conshohocken, PA: 2013.
18. Schulte, U., Weihmann, J. and Mansfeldt, T., Optimized enrichment and purification of ferrocyanide for 13/12C and 15/14N isotope analysis of aqueous solutions. *Water Research*, 2010, **44**(18), 5414-5422.
19. IULTCS, Sampling. *J. Soc. Leather Technol. Chem.*, 2000, **84**(7), 303.
20. IUP6, Measurement of tensile strength and percentage elongation. *J. Soc. Leather Technol. Chem.*, 2000, **84**(7), 317.
21. IUP8, Measurement of tear load. *J. Soc. Leather Technol. Chem.*, 2000, **84**(7), 317.
22. Shi, B., He, X. Q. and Haslam, E., Gelatin-polyphenol interaction. *J. Amer. Leather Chem. Ass.*, 1994, **89**(7), 98.
23. Gunter, R., From collagen to leather – the theoretical background. Ludwigshafen: BASF service center media and communications, 2007: 24.
24. Covington, A. D., Modern tanning chemistry. *Chemical Society Reviews*, 1997: 111-126.
25. Nie, J., Du, B. and Oppermann, W., Swelling, elasticity, and spatial inhomogeneity of poly (N-isopropylacrylamide)/clay nanocomposite hydrogels. *Macromolecules*, 2005, **38**(13), 5729-5736.
26. Okay, O. and Oppermann, W., Polyacrylamide-clay nanocomposite hydrogels: rheological and light scattering characterization. *Macromolecules*, 2007, **40**(9), 3378-3387.
27. Shi, B. and Di, Y., Plant polyphenols. Beijing: Science Press, 2000, 144.
28. Saravanabhavan, S., Thanikaivelan, P., Rao, J. R. et al., Natural leathers from natural materials: progressing toward a new arena in leather processing. *Environmental Science & Technology*, 2004, **38**(3), 871-879.
29. Essawy, H. A., Abd El-Wahab, N. A. and Abd El-Ghaffar, M. A., PVC-laponite nanocomposites: Enhanced resistance to UV radiation. *Polymer Degradation and Stability*, 2008, **93**, 1472-1478.

A Novel Approach for Lightfast Wet-white Leather Manufacture Based on Sulfone Syntan-aluminum Tanning Agent Combination Tannage

by

Long Zhang,[1] Xinyu Zhao, Chunhua Wang[2]* and Wei Lin[1]

[1]Key Laboratory of Leather Chemistry and Engineering of Ministry of Education, Sichuan University,
Chengdu, China, 610065

[2]Department of Biomass and Leather Engineering, College of Light Industry, Textile and Food Engineering,
Sichuan University, Chengdu, China, 610065

Abstract

Wet-white tanning as an eco-friendly leather-making process has been attracting considerable attention. Herein, we have investigated a novel combination tannage for lightfast wet-white leather based on sulfone syntan and aluminium tanning agent. By optimizing the technology, 10% sulfone syntan and 3% aluminum tanning agent at final pH 4.0 - 4.5 can raise the shrinkage temperature (T_s) of the wet-white leather to ~81ºC. The synergistic tanning mechanism of the two has been illustrated. As verified by Zeta potential measurements, the introduction of Al^{3+} into the sulfone syntan system led to the increase in the isoelectric point (IEP) of wet-white leather, which is favorable for the subsequent post-tanning process. Scanning electron microscope-Energy dispersive X-ray spectroscopy (SEM-EDX) results reveal that sulfone syntan and aluminum tanning agent can be evenly bound within the leather matrix and promote the formation of tightly woven networks of collagen fibers. The novel combination tanning approach not only improves light fastness and lighter shade, but also confers high physical and mechanical properties to the wet-white leather.

Introduction

Chrome tanning is the most widely used tanning method for leather making worldwide because of its various advantages.[1] However, it also causes negative effects on the environment due to the pollution of tannery effluents and improper disposal of chrome-containing solid wastes.[2] With the rising of eco-environmental awareness, wet-white or chrome-free tanning as an eco-friendly technology has attracted increasing attention.[3] Nowadays, there is a globally growing requirement for wet-white leather products, such as automotive leather, upholstery leather and garment leather.[4,5,6] Besides, pastel shade leathers are popular in today's fashion world.[7] So far, there are many reports concerning the use of white minerals such as aluminum (III), zirconium (IV) and titanium (IV) salts, vegetable tanni,[8] syntan,[9] aldehydes and silica,[10] to replace or reduce using conventional chrome tanning materials. Among them, vegetable tanning agent is considered as a promising alternative owing to its natural origin and appropriate tanning properties.[11] Nevertheless, it also faces several constraints like poor permeability, weak light fastness and difficulty in making pastel shades.[12]

Sulfone syntan is a type of small molecule alternative syntan with sulfone bridges deriving from phenol compounds. Owing to its good water dispersibility, permeability and absorbability, especially excellent light fastness and yellowing resistance, it has been developed as a potential tanning agent for wet-white leather manufacture.[13] However, sulfone syntan exhibits disadvantages of insufficient strength, strong surface negative charge and limited hydrothermal stability.[14] It is noted that the combination tannage, like classic vegetable-aluminum combination tannage, is an very effective approach to achieve complementary characteristics. Aluminum-mimosa combination tannage can give wet-white leathers with shrinkage temperature about 100ºC.[12] Considering the structural similarity between vegetable tanning agent and sulfone syntan, there might be a certain synergistic tanning effect between sulfone syntan and aluminum tanning agent.[15] In addition, the introduction of electropositive aluminum tanning agent to sulfone syntan tannage, is also expected to improve physical strength performance and reduce the surface negative charges, thus benefiting the post-tanning procedures for light-colored leather manufacturing.[12, 16, 17]

Recently, we have optimized solo sulfone syntan tanning process.[14] In the present work, a novel sulfone syntan-Al^{3+}

---

*Corresponding authors e-mail: wangchunhua@scu.edu.cn; wlin@scu.edu.cn
Manuscript received March 19, 2018, accepted for publication April 9, 2018.

combination tanning system has been explored toward lightfast wet-white leather making. Our aim is to develop feasible chrome-free tanning technologies to counter ecological constraints and stricter requirements for leather performance properties.

Experimental

Materials

The pickled goatskins with an average area of 4 - 5 square meters were used for tanning processing. The dosages of the chemicals were all based on the weight of pickled pelts thereafter. Aluminum tanning agent, BN was a courtesy of BASF, Germany. Sulfone syntan, BC was purchased from Silvateam Co. Ltd., Italy. All chemicals used for leather processing were of commercial grade and the others were of analytical grade.

Wet-white Leather Tanning Processes

The BC and BN solo tannages were first optimized in our laboratory before their combination tanning. The details of solo BC tanning process can be found in our previous work.[14] For solo BN tanning, the pickled pelts were re-pickled with HCOOH solution (1.5%, w/w) to adjust the pH of pelts to ~3.0 for 1 h. Then BN of 1 - 5% was respectively added to determine its appropriate amounts. After the drum was run for 4 h, the tanning floats were basified to pH ~3.5 using NaHCO$_3$ solution (1.5%, w/w). After another 120 min of running, the drum stopped and stayed overnight.

During BC-BN combination tanning process, the tanning with BC was started by de-pickling the pickled hides with NaHCO$_3$ solution (1.5%, w/w) to raise the pH of the pelts to ~4.5. Then 7.5%, 10% or 15% of BC was added under constant running for 4 h, respectively. Subsequently, the pH of the tanning float was adjusted to ~3.0 and BN of 1 - 5% was further added for 6 h running to fulfil the combination tanning. The final pH of tanning float was adjusted to 3.0 - 5.5 with HCOOH solution or NaHCO$_3$ solution, respectively. After another 120 min of running, the drum stopped overnight. The resulted wet-white leathers were washed with 200% of water and piled for 24 h. T_s of the leathers were then measured by a shrinkage tester, and the tanning floats were sampled for the determination of the residual BN.

Determination of Shrinkage Temperature

Shrinkage temperature (T_s) was measured by a shrinkage tester using ASTM method.[18] A 10 mm × 60 mm specimen was cut out from each leather sample and was held in water, heated at a rate of 2°C per minute. The temperature at the first definite sign of shrinking was noted as T_s. Each test was done in triplicate.

ICP-AES Measurements

The content of BN in the wet-white leather and its residue in tanning floats was measured with inductively coupled plasma-atomic emission spectrometry (ICP-AES) technique by monitoring the Al trace as the characteristic element of BN.[19] 0.1 g wet-white leather samples or 2 mL spent tanning floats were digested by 10 mL HCl/HNO$_3$ solution (v/v, 3:1) for 2 h at 110°C, respectively. The digested solution was filtered through Millipore hydrophilic GTTP04700 membrane with a 0.22 μm pore size, and the filtrate was diluted to 100 mL with deionized water. Then 5 mL diluted filtrate was taken for the measurements of Al^{3+} concentration by an ICP-AES spectrometer (Optima 2100DV, Perkin-Elmer, USA). The analytical wavelength for Al was 396.152 nm. The uptake of BN and its residue in tanning floats were calculated. Each set of test was done in triplicate.

Zeta Potential Measurements

The Zeta potentials of the wet-white leather samples at different pH were measured using a flow potential analyzer (Mütek™SZP-10, BTG, Germany). 10 g samples were pulverized into particles with the size of less than 308 mm, and then dispersed in 400 mL deionized water. The pH of the suspension was adjusted ranging from 3.0 to 10.0 with an appropriate amount of 0.1 M HCl or NaOH solution, respectively. The obtained suspension was shaken at 150 rpm for 30 min at 30°C. Next, the suspension was pumped into the measuring chamber under vacuum conditions for the measurement.

SEM-EDX Analysis

After lyophilization at -43°C in a freeze dryer (Alpha 1-2 LD, Christ, Germany) for 24 h, the wet-white leather samples were cut into the specimens with a thickness of 1.0 mm by a microtome (CM1900, Leica, Germany) before observations. The morphologies of leather specimens were recorded by a scanning electron microscope (JSM-7500F, JEOL, Japan) at an accelerating voltage of 15 kV different lower and higher magnification levels. The corresponding elemental compositions of the specimens were confirmed by a coupled energy dispersive X-ray spectroscopy (EDX) detector.

Color Measurements

Wet-white leather samples were radiated by 15 W UV lamps at room temperature. Color parameters of leather samples with different irradiation time were measured by Chromaticity analyzer (8200, X-RITE, USA) at the spectral detection wavelength of 400 ~ 700 nm. The color values are lightness-darkness (L^*), redness-greenness (a^*) and yellowness-blueness (b^*) calculated by the spectrophotometer coupling an analytical software (UV-2401PC Color Analysis Software, Shimadzu, Japan). The color difference ΔE was calculated by the CIE 1976 LAB $L^*a^*b^*$ color space, as the following equations:

$$\Delta E = \sqrt{(\Delta L *)^2 + (\Delta a *)^2 + (\Delta b *)^2}$$

Where ΔL^*, Δa^* and Δb^* represent the lightness, redness-greenness and yellowness-blueness differences of wet-white leather in comparison with no UV irradiation wet-white leather, respectively. ΔE represents the total color difference. Each test was done in triplicate.

Test of Physical and Mechanical Properties
Wet-white leather specimens were sampled according to the approaches recommended by the International Union of Leather Technologists and Chemists Societies (IULTCS).[20] Physical and mechanical properties, such as tensile strength, tear strength, elongation and bursting strength, were tested as per the standard procedures.[21-23] Each set of tests was done in triplicate.

Results and Discussion

BC-BN Combination Tannage
As can be seen from Figure 1a, the solo BN tanning at 3 - 5% dosage raises the shrinkage temperature (T_s) of pickled hides by ~8°C. The solo BC tanning at 7.5 - 15% can typically enhance the T_s to 65 - 69°C, increased by ~15°C at most, consistent with previous report[14]. In contrast, the T_s of the BC-BN combination tanned leather reaches to ~81°C, which is higher than the sum of the individual contributions of BC and BN to the T. Moreover, it is clear that the uptake of BN (Figure 1b) markedly increase with increasing dosage of BC in the combination tanning, implying that BC can facilitate the fixation of BN in the wet-white leather. Therefore, there is a certain synergistic tanning effect between sulfone syntan, BC and Al^{3+} from BN in the combination tanning process.

Note that when the dosage of BN increases from 3% to 5%, the T_s of the combination tanned leather increases slightly, but its uptake does not increase anymore, and even decreases instead. This can be related to the blockage of available bonding sites of collagen fibers by the BC and BN at certain concentrations. Therefore, 10% BC and 3% BN are used in the subsequent combination tanning experiments, which can confer the wet-white leather with shrinkage temperature of ~81°C.

The synergistic tanning mechanism of BC and BN is graphically illustrated in Figure 2. It is speculated that the Al^{3+} of BN can not only react with carboxyl groups of collagen, but also can interact with phenolic hydroxyl groups of BC molecules via complexation bonding.[15] In addition, the coordinated BC-BN complex can also create bridges between the collagen fibers, thus leading to the formation of cross-linked network in the hide matrix and the increase in T_s.[24-26]

Figure 3 shows the effect of final pH on T_s of wet-white leathers. It is clear that pH ranging from 4.0 - 4.5 can mostly benefit the improvement in T_s. The reason is connected with reactive functional groups of collagen involved in the tanning process, and also with the synergistic action of BC and BN. Therefore, the final pH of 4.0 - 4.5 is preferred in the combination tannage, indicating stronger combination tanning effects of BC and BN. Therefore, the final pH and the BN tanning time in the combination tannage are optimized as pH 4.0 - 4.5 and 3 h, respectively.

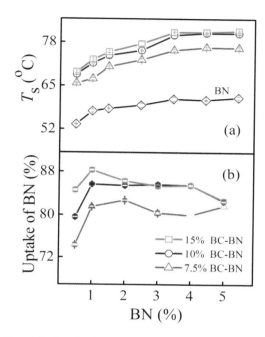

Figure 1. T_s of wet-white leathers (a) and uptake of BN (b) as a function of BN dosage.

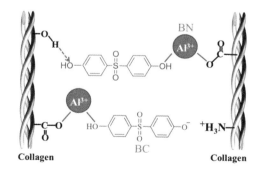

Figure 2. Schematic representation showing the synergistic crosslinking of BC and BN with collagen in the combination tannage.

Zeta Potential Analysis

The isoelectric point (IEP) of collagen or leather can be obtained by the measurement of the zeta potential, and the zero point indicates the IEP.[24] Figure 4 shows that the IEP of solo BC tanned leather is 3.12; whereas for BC-BN combination tanned leather, the IEP shifts to 4.73. This is apparently owing to the influence of tanning materials. In the former case, blockage of the NH_3^+ groups by BC leads to lower IEP than that of pickled hides (IEP ~5.3). And in the latter case, the coordination bonding between Al^{3+} from BN and the carboxyl groups in the side chain of collagen molecules, and also phenolic hydroxyl group in BC molecule result in the increase of IEP.[25-28] Additionally, it reveals that when the pH of the float is 4.0 - 4.5 (< IEP ~4.73), the surface charge of the combination tanned leather are positively charged, which is favorable for a high uptake of anionic retanning chemicals, fatliquors and dyes in the following processes.[29]

SEM-EDX Analysis

SEM analysis was carried out to study the influence of the combination tanning on morphological characteristics of wet-white leathers. As can be seen from Figure 5a, solo BC tanned collagen fibers are closely woven with each other to form a stratified collagen fiber bundles. At a higher magnification level (Figure 5b), it shows

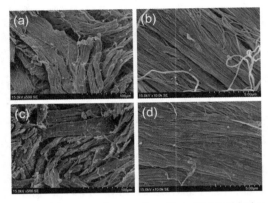

Figure 5. SEM images of cross sections of solo BC tanned leather (a and b) and BC-BN combination tanned leather (c and d).

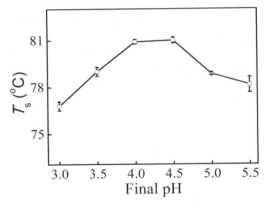

Figure 3. T_s of combination tanned leathers as a function of final pH.

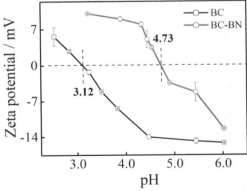

Figure 4. Effect of pH on the Zeta potential of wet-white leathers.

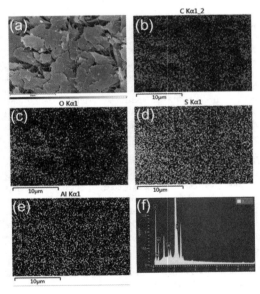

Figure 6. SEM image of vertical section of the BC-BN combination tanned leather (a); EDX elemental mappings of carbon (b), oxygen (c), sulfur (d) and aluminum (e); and corresponding EDX spectrum (f).

clear fibrils with smooth surface. After the introduction of BN (Figure 5c and d), the combination tanned collagen fibers are more tightly woven due to the effect of Al^{3+} tanning agent.

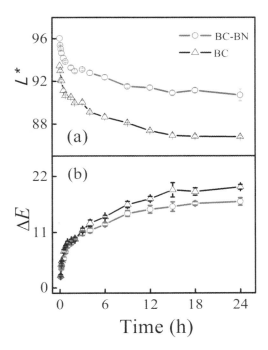

Figure 7. L' (a) and ΔE (b) of wet-white leathers under UV radiation for 24 h.

Table I
Physical and mechanical properties of BC-BN combination tanned leathers.

Parameters	BC-BN tanned leathers	Standard for shoe upper leather[31]
Tensile strength (MPa)	17.2±1.3	--
Tear strength (N)	40.0±1.6	≥20
Elongation at 10N (%)	16.9±2.4	≤40
Bursting strength (N/mm)	207.4±3.8	≥200
T_s (°C)	80.9±0.4	≥80

The distribution of tanning agents in collagen fibers was also studied as shown in the corresponding EDX spectra (Figure 6). Both sulfur (S) and aluminum (Al), the feature elements in BC and BN, respectively, are homogeneously distributed in the vertical section of the wet-white leather, demonstrating that BC and BN can uniformly penetrate into leather matrix. It can be expected that a reasonable good wet-white leather properties can be resulted.

Color Analysis

Compared with solo BC tanning, the BC-BN combination tanned leather gives a lighter shade, as indicated in the L' value (Figure 7a). Light fastness is assessed by the exposure of the wet-white leathers in UV irradiation.[30] It can be seen that the L' value of the combination tanned leather is always higher, whereas the color difference (ΔE) is lower (Figure 7b) for an extended irradiation time, suggesting that the introduction of BN can endow wet-white leathers with better light fastness. The result shows that BC-BN combination tannage is favorable for light-colored leather manufacturing.

Physical and Mechanical Properties Measurements

Table I presents physical and mechanical properties of the wet-white leather tested with the official methods recommended by IULTCS. Note that the tear strength and bursting strength of the obtained leather can reach up to 40 N and 207 MPa, respectively. All of these physical and mechanical properties can meet the Chinese standard requirements for shoe upper leather.[31] It suggests that the BC-BN combination tannage is a promising approach and can be applied in shoe upper leather manufacture.

Conclusions

In the present study, a novel combination tanning approach based on BC and BN for lightfast wet-white leather manufacture has been established. The results show that sulfone syntan and aluminum tanning agent exhibit synergistic tanning effect in the combination tannage. The 10 % of BC and 3% BN at final pH 4.0 - 4.5 can endow the leather with T_s above 80°C. Both of the two are evenly bound within the the leather matrix and make the collagen fibers isolated and compact. The introduction of BN (Al^{3+}) into BC (sulfone syntan) increases positive electrical property of the wet-white leather and gives it lighter shade and better light fastness. The combination tanned leathers have reasonable good physical and mechanical properties that can meet official standard requirements for shoe upper leather.

Acknowledgements

The financial support of National Natural Science Foundation (NNSF) of China (21476148), Innovation Team Program of Science & Technology Department of Sichuan Province (Grant 2017TD0010), and National Key R&D Program of China (2017YFB0308402) are gratefully acknowledged.

References

1. Zhang, C. X., Xia, F. M., Long, J. J. and Peng, B. Y.; An integrated technology to minimize the pollution of chromium in wet-end process of leather manufacture. *J Clean. Prod.* **154**, 276-283, 2017.
2. Bacardit, A., van der Burgh, S., Armengol, J. and Ollé, L.; Evaluation of a new environment friendly tanning process. *J Clean. Prod.* **65**, 568-573, 2014.
3. Krishnamoorthy, G., Sadulla, S., Sehgal, P. K. and Mandal, A. B.; Greener approach to leather tanning process: d-Lysine aldehyde as novel tanning agent for chrome-free tanning. *J Clean. Prod.* **42**, 277-286, 2013.
4. Taylor, M.M., Lee, J., Bumanlag, L. P., Balada, E. H. and Brown, E. M.; Treatments to enhance properties of chrome-free (wet white) leather. *JALCA* **106**, 35-43, 2011.
5. Wolf, G., Breth, M., Carle, J. and Igl, G.; New developments in wet white tanning technology. *JALCA* **96**, 111-119, 2001.
6. Shi, J. B., Ren, K. S., Wang, C. H., Wang, J. and Lin, W.; A novel approach for wet-white leather manufacture based on tannic acid-laponite nanoclay combination tannage. *J. Soc. Leather Technol. Chem.* **100**, 25-30, 2016.
7. Fathima, N. N., Kumar, T. P., Kumar, D. R., Rao, J. R. and Nair, B. U.; Wet white leather processing: a new combination tanning system. *JALCA* **101**, 58-65, 2006.
8. Sivakumar, V., Princess, A., Veena, C., Devi, R. L.; Ultrasound Assisted Vegetable Tannin Extraction from Myrobalan (Terminalia Chebula) Nuts for Leather Application. *JALCA* **113**, 53-58, 2018.
9. Olle, L., Jorba, M., Font, J., Shendrik, A. and Bacardit, A.; Biodegradation of wet-white leather. *J. Soc. Leather Technol. Chem.* **95**, 116-120, 2011.
10. Liu, Y. S., Chen, Y., Yao, J., Fan, H. J., Shi, B. and Peng, B. Y.; An environmentally friendly leather-making process based on silica chemistry. *JALCA* **105**, 84-93, 2010.
11. Shi, J. B., Puig, R., Sang, J.and Lin, W.; A comprehensive evaluation of physical and environmental performances for wet-white leather manufacture. *J Clean. Prod.* **139**, 1512-1519, 2016.
12. Fathima, N. N., Saravanabhavan, S., Rao, J. R. and Nair, B. U.; An eco-benign tanning system using aluminiuum, tannic acid, and silica combination. *JALCA* **99**, 73-81, 2004.
13. Feng, G. T., Shan, Z. H., Shao, S. X., and Jiang, L.; Preparation and tanning powers of syntan with sulphone. *Leather Science and Engineering* **13**, 50-53, 2003.
14. Li, X. P., Sang, J., Shi, J. B. and Lin, W.; Study on the tanning property of soufone syntan tanning agents. *China Leather* **45**, 18-22, 2016.
15. Madhan, B., Aravindhan, R., Siva, M. S., Sadulla, S., Rao, J. R. and Nair, B. U.; Interaction of aluminum and hydrolysable tannin polyphenols: An approach to understanging the mechanism of aluminum vegetable combination tannage. *JALCA* **101**, 317-323, 2006.
16. Covington, A. D. and Sykes, R. L.; The use of aluminum salts in tanning. *JALCA* **79**, 72-93, 1984.
17. Bacardit, A., Gonzalez, M., Van der Burgh, S., Armengol, J. and Olle, L.; Development of a new leather intermedidate: wet-bright with a high dye affinity. *JALCA* **111**, 113-122, 2016.
18. ASTM International. Annual Book of ASTM Standard, West Conshohocken, PA: 2013.
19. Metreveli, G., Kaulisch, E. M. and Frimmel, F. H.; Coupling of a column system with ICP-MS for the characterisation of colloid-mediated metal (loid) transport in porous media. *Acta Hydrochim. Hydrobiol.* **33**, 337-345, 2005.
20. IULTCS, Sampling. *J. Soc. Leather Technol. Chem.* **84**, 7 (extra), 303-309, 2000.
21. IUP 6, Measurement of tensile strength and percentage elongation. *J. Soc. Leather Technol. Chem.* **84**, 7 (extra), 317-321, 2000.
22. IUP 8, Measurement of tear load - double edge tear. *J. Soc. Leather Technol. Chem.* **84**, 7 (extra), 327-329, 2000.
23. SLP 9 (IUP 9), Measurement of distension and strength of grain by the ball burst test. Official methods of analysis. The Society of Leather Technologists and Chemists, Northampton, 1996.
24. Reich, G.; From collagen to leather: The theoretical background. Ludwigshafen: BASF Service Center Media and Communications. 179-180, 2007.
25. Sykes, R. L. and Cater, C. W.; Tannage with aluminum salts.1. Reactions involving simple polyphenolic compounds. *J. Soc. Leather Technol. Chem.* **64**, 29-31, 1980.
26. Sykes, R. L., Hancock, R. A. and Orszulik, S. T.; Tannage with aluminum salts.2. Chemical basis of the reactions with polyphenols. *J. Soc. Leather Technol. Chem.* **64**, 32-37, 1980.
27. Brown, E. M. and Dudley, R. L.; Approach to a tanning mechanism: Study of the interaction of aluminium sulfate with collagen. *JALCA* **100**, 401-409, 2005.
28. Andrade, Â. L., Ferreira, J. M. F. and Domingues, R. Z.; Zeta potential measurement in bioactive collagen. *Materials Research* **7**, 631-634, 2004.
29. Matsumoto, S., Ohtaki, A. and Hori, K.; Carbon fiber as an excellent support material for wastewater treatment biofilms. *Environ. Sci. Technol.* **46**, 10175-10181, 2012.
30. Ozgunay H.; Lightfastness properties of leathers tanned with various vegetable tannins. *JALCA* **103**, 345-351, 2008.
31. Chen, R., H., Chen, R., S.; QB/T 1873-2010: shoe upper leather, *China Light Industry Press*, Beijing China, pp. 1-4, 2010.

Appendix III

LANGMUIR

Article

pubs.acs.org/Langmuir

Diffusion and Binding of Laponite Clay Nanoparticles into Collagen Fibers for the Formation of Leather Matrix

Jiabo Shi,[†,§] Chunhua Wang,[†] To Ngai,*[,‡] and Wei Lin*[,†]

[†]National Engineering Laboratory for Clean Technology of Leather Processing, Sichuan University, Chengdu 610065, China
[‡]Department of Chemistry, The Chinese University of Hong Kong, Shatin, N. T., Hong Kong, China
[§]College of Bioresources Chemical and Materials Engineering, Shaanxi University of Science and Technology, Xi'an 710021, China

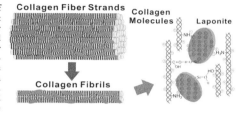

ABSTRACT: Understanding accessibility and interactions of clay nanoparticles with collagen fibers is an important fundamental issue for the conversion of collagen to leather matrix. In this study, we have investigated the diffusion and binding of Laponite into the collagen fiber network. Our results indicate that the diffusion behaviors of Laponite into the collagen exhibit the Langmuir adsorption, verifying its affinity for collagen. The introduction of Laponite leads to a shift in the isoelectric point of collagen from ~6.8 to ~4.5, indicating the ionic bonding between the positively charged amino groups of the collagen and negatively charged Laponite under the tanning conditions. Fluorescence microscopy, atomic force microscopy, field-emission scanning electron microscopy, energy-dispersive X-ray spectroscopy, and wide-angle X-ray diffraction analyses reveal that Laponite nanoparticles can penetrate into collagen microstructure and evenly distributed onto collagen fibrils, not altering native D-periodic banding patterns of collagen fibrils. Attenuated total reflectance-Fourier transform infrared and Raman spectroscopy detections further demonstrate the presence of noncovalent interactions, namely, ionic and hydrogen bonding, between Laponite and collagen. These findings provide a theoretical basis for the use of Laponite as an emerging tanning agent in leather manufacture.

■ INTRODUCTION

Collagen is a primary structural component of mammalian connective tissues, including skin, tendon, and bone, and provides main mechanical support and structural integrity for living organisms.[1] The main protein in skin is fibril-forming type I collagen, and its commercial and industrial significance is exemplified by traditional leather industry and current biomedical applications.[2,3] The hierarchical structure of collagen from the triple helix (~300 nm in length and ~1.5 nm in diameter) to microfibril (~40 nm in diameter), fibril (100−200 nm in diameter), and further collagen fiber has been well documented.[4] A key procedure in the transformation of collagen into leather matrix is tanning. In the process, collagen reacts with a variety of active substances, such as mineral salts,[5] vegetable tannins,[6] aldehydes,[7] nanosilicates,[8] and SiO_2,[9] resulting in the changes in appearance and properties. The interactions and thus structural modification to collagen fibers are responsible for the formation of leather matrix; nevertheless, the crucial premise is their accessibility, i.e., suitable dimensions of diffusion paths and available specific surface area of collagen. Therefore, the diffusion and penetration of active substances at the actual centers of reaction and bonding of collagen fibers must be considered.

Laponite is a uniform synthetic disk-shaped hectorite clay with a thickness of 1 nm, a diameter of 25−30 nm, and a negative surface charge when suspended in water.[10] It has been widely used in various formulations of waterborne products, like industrial and household coatings, cleansers, and personal care products, due to its nontoxicity, excellent colloidal property, and biocompatibility.[11] Also, it has been applied as stabilizers in Pickering emulsion polymerization[10,12] and as multifunctional inorganic cross-linkers to synthesize novel nanocomposite hydrogels[13] and gelatin−clay nanocomposite films with improved physical properties.[14] Moreover, compared with other clay particles, such as montmorillonite[15] and attapulgite,[16] Laponite is more promising as an alternative to conventional chrome tanning agent for eco-leather manufacture due to its much smaller particle size and conceivably better penetration inside hide collagen fibers.

In our previous studies, we have developed a novel tannic acid/Laponite combination tanning method for the manufacture of wet-white leather with improved environmental benefits.[17] As a result, the introduction of Laponite gains a marked improvement in the hydrothermal stability (denoted as shrinkage temperature, T_s) and the mechanical strength over solo tannic acid-tanned leather, showing the synergistic tanning effect.[18] It is evidenced that vegetable tannins react with collagen primarily via hydrogen bonding, occurring at various structural hierarchies.[19,20] However, there is relatively little knowledge about the accessibility and the interactions of

Received: March 21, 2018
Revised: April 26, 2018
Published: May 28, 2018

171

Laponite within hide collagen fibers. This is distinctly different from the fabrication of collagen/clay nanocomposite biomaterial that used acid soluble collagen,[3] instead of natural collagen fiber network.

In this study, we have investigated the diffusion behaviors of Laponite into the collagen fibers and its influences on the hierarchical structure of collagen. Our aim is to explore the accessible surface of the collagen structural elements (i.e., molecule, microfibril, fibril, or fiber) where Laponite can reach, because it influences the further binding, which is of significant for the conversion of collagen fiber to leather.

■ EXPERIMENTAL SECTION

Materials. Raw collagen fibers were taken from fresh healthy cattle hides, and the beamhouse process was performed under a condition for commercial leather making, similar to the previous report.[21] Laponite RDS, $Na^{+0.7}[(Si_8Mg_{5.5}Li_{0.3})O_{20}(OH)_4]^{-0.7}$, modified with pyrophosphate ($P_2O_7^{4-}$) was purchased from BYK Additives & Instruments Ltd. (U.K.). Its chemical composition (dry basis) is SiO_2 (54.5 wt %), MgO (26.0 wt %), Li_2O (0.8 wt %), Na_2O (5.6 wt %), and P_2O_5 (4.4 wt %), provided by the supplier and often cited in research articles.[10] Fluorescent laser dye rhodamine 6G was obtained from Aladdin Chemistry Co. Ltd. (China).

Sample Preparation. The obtained pure collagen fiber or pelt after beamhouse processing was weighted as the base of dosage. It was first immersed in $NaCl/H_2SO_4$ solution at pH 2.5 for 30 min, next the pH of the float was adjusted to 5.0 with $NaHCO_3$ solution (1.5%, w/w) for the Laponite tanning.[18] And then different amounts of Laponite were added under constant stirring for 4 h so as to sufficient diffusion/penetration into hide collagen. The initial concentrations of Laponite (C_i) were 10.0–60.0 mg/L. Subsequently, the pH was adjusted to 3.0 by adding HCOOH solution (1:10, v/v). After the tanning reaction for 24 h, the resultant wet-white leathers were rinsed with deionized water to remove unfixed Laponite.

Adsorption Capacity Measured by Inductively Coupled Plasma-Atomic Emission Spectrometry (ICP-AES). The adsorption capacity of Laponite on collagen was determined with inductively coupled plasma-atomic emission spectrometry (ICP-AES) by measuring the Mg trace in Laponite on a PerkinElmer Optima 2100DV spectrometer according to the reported approach.[22] The concentrations of Laponite in the residual floats were then obtained by mass balance calculation. The results were the average of three measurements.

ζ Potential Measurements. ζ potential was determined by streaming potential method using a BTG Mütek SZP-10 analyzer according to the reported method.[23] The wet-white leathers were first pulverized into powders and then dispersed in deionized water. The pH of the suspension was adjusted with 0.1 M HCl or NaOH solution. The obtained suspensions were vibrated at 150 rpm for 30 min at 25 °C and then pumped into the measuring cell under vacuum conditions.

Fluorescence Microscopy Observations. The fluorescent labeling of Laponite was according to the reported approach.[24] For the observation, the labeled Laponite was specially used to treat collagen fibers under the same tanning conditions as above. The frozen slices with a thickness of 50 μm were prepared by a Leica CM1900 microtome. Fluorescence microscopy images of the collagen fibers were recorded on a Nikon Ti-U inverted microscope coupled with a charged coupled device camera, a mercury lamp (130 W), and a set of optic filters, including filters (340–380 nm) for UV light excitation.

Field-Emission Scanning Electron Microscopy (FE-SEM)/Energy-Dispersive X-ray (EDX) Spectroscopy Analysis. The surface microstructures of collagen fibers and leather matrixes were observed on a JEOL JSM-7500F field-emission scanning electron microscope at an accelerating voltage of 15.0 kV after sputter coated with gold. The corresponding relative elemental mappings and compositions were also performed by a coupled energy-dispersive X-ray (EDX) spectroscopy detector with an acceleration voltage of 20.0 kV.

Atomic Force Microscopy (AFM) Observation. Topographic images were acquired on a Shimadzu SPM-9600 atomic force microscope under tapping mode. The frozen slices with a thickness of 50 μm were dried at room temperature for 24 h in a desiccator with silica gel to avoid possible contamination just before observation.

Wide-Angle X-ray Diffraction (WAXD) Studies. Wide-angle X-ray diffraction (WAXD) measurements were performed on a Philips Analytical X'pert ProMPD X-ray diffractometer coupled with a plumbaginous-monochromated Cu Kα irradiation of 0.154 nm working at 30 kV and 35 mA. The specimens were scanned in the diffraction angle (2θ) ranging from 5 to 60° with a scanning rate of 2°/min and a step size of 0.03° at ambient temperature and humidity. The real lattice space d that represents characteristic structure dimension can be calculated by $d = \lambda/(2 \sin \theta)$, where λ is the X-ray wavelength and θ is the half of diffraction angle.[21]

Attenuated Total Reflectance-Fourier Transform Infrared (ATR-FTIR) Analysis. Attenuated total reflectance-Fourier transform infrared (ATR-FTIR) spectra were recorded on a Nicolet iS10 FTIR spectrometer with a Thermo DTGS detector and an attenuated total reflection accessory with a Thermo iD5 ZnSe crystal reflection element at room temperature. All spectra were obtained with a resolution of 4 cm^{-1} over the range of 650–4000 cm^{-1}. The spectral plots represent the average of 32 scans.

Raman Spectral Measurements. Raman scattering measurements were carried out on a confocal Horiba Labra HR Raman spectrometer excited by a laser at 785 nm and collected by an Olympus objective (100×, 0.90 NA). The output power of the laser was fixed to 20 mW, and the laser power on the sample surface is about 5 mW. The Raman shifts were carefully calibrated using Si plate with an uncertainty of 0.5 cm^{-1}. The Raman spectra were corrected for biological fluorescence by subtraction of a fourth-order polynomial, which was fit to the spectrum. Spectral intensity was normalized using C–C vibrational mode of phenylalanine aromatic ring (1004 cm^{-1}) as an internal standard.

■ RESULTS AND DISCUSSION

The conversion of hide into leather is a multistage process of physical and structural modification of the collagen. In the

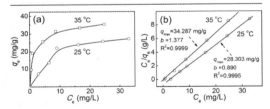

Figure 1. (a) Adsorption capacity and (b) Langmuir isotherms of Laponite onto collagen fibers.

Table 1. Relations between the Concentrations of Laponite in the Residual Floats at Equilibrium (C_e) and Its Initial Concentrations (C_i)

	C_e (mg/L)	
C_i (mg/L)	25 °C	35 °C
10.0	2.9	0.5
20.0	6.49	0.9
30.0	9.09	4.4
40.0	16.2	8.7
50.0	24.4	16.7
60.0	33.0	24.8

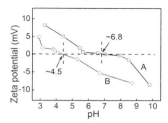

Figure 2. ζ potential of collagen fibers (A) before and (B) after addition of Laponite at an initial concentration of 40 mg/L.

beamhouse, noncollagenous components, such as keratins, proteoglycans, and fats, are removed, which is necessary to allow fiber structure to loosen. This "opening-up" effect promotes accessibility to the reactive centers in the microstructure of the collagen.

The diffusion and penetration of Laponite within collagen structure as well as its affinity for collagen was evaluated by the adsorption capacity of Laponite on the collagen during the tanning process (Figure 1). The relations between the concentrations of Laponite in the residual floats at equilibrium (C_e) and its initial concentrations (C_i) are given in Table 1. The data in Figure 1a and Table 1 show that the adsorption capacity increases with increasing C_i and reaction temperature; whereas, the adsorption approaches to a plateau of the amount adsorbed of Laponite at equilibrium state (q_e) when C_i is above 40 mg/g. In addition, Laponite may self-aggregate to form larger particles at high concentration in the floats,[25] blocking their diffusion into the fibers. It is known that the simplest diffusion, e.g., the substances with no affinity for collagen in a static system, is triggered by the concentration gradient and follows Fick's laws.[26] Otherwise, the situation is different when the diffusion is considerably accelerated by constant affinity adsorption. To further understand the adsorption mechanism of Laponite on collagen fibers, the adsorption capacity was analyzed with respect to the Langmuir model.[27,28] It can be seen that the experimental data fit well with the Langmuir model (Figure 1b), and the equation can be expressed as follows.[29]

$$\frac{C_e}{q_e} = \frac{1}{q_{max}b} + \frac{C_e}{q_{max}} \quad (1)$$

where the maximum adsorption capacity of Laponite (q_{max}, mg/g) and the adsorption constant (b) were calculated from the slope and intercept, respectively. Here, high R^2 values indicate that the model describes the adsorption behavior well. Meanwhile, a strong affinity exists for Laponite onto the collagen. Note that there are higher q_{max} and b values at 35 °C than those at 25 °C, implying the endothermic adsorption.

Figure 2 shows that the isoelectric point (IEP) of Laponite-tanned leather decreases from ∼6.8 to ∼4.5 due to the introduction of negatively charged Laponite. Considering the primary structure of collagen, it can be speculated that the forces triggering and promoting diffusion and penetration of Laponite must involve the electrostatic attraction from the positively charged amino groups of the collagen.[30]

The diffusion of Laponite into the collagen fibers with time is visualized by fluorescence microscopy, as shown in Figure 3. It is clear that, after treatment with the labeled Laponite, the collagen fibers exhibit red fluorescence, indicating the adsorption and binding site of Laponite. The fluorescence images also show that Laponite has penetrated into collagen microstructure owing to its favorable nanodispersibility.[14,17] Furthermore, the nanoclay particles predominately deposit on the surface of collagen fibrils as reflected in the highlighted fibrillar outlines (Figure 3b–d), showing its high affinity for collagen.

To study the influence of Laponite on the hierarchical structure of the collagen, we combined the results from FE-SEM and AFM in Figure 4. As can be seen in Figure 4a, pure collagen fibers exhibit interwoven flexuous fiber strands formed by collagen fibrils. After addition of Laponite, the formed large clay platelets with diameters from nanometers to submicrons

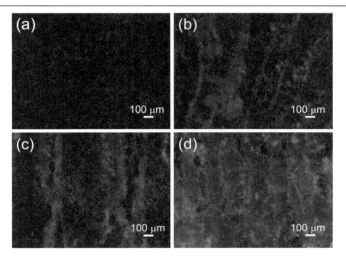

Figure 3. Fluorescence microscopy images of collagen fibers after addition of Laponite stained with rhodamine 6G: (a) 0 h, (b) 0.5 h, (c) 1.0 h, and (d) 2.0 h.

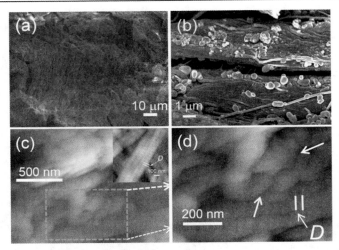

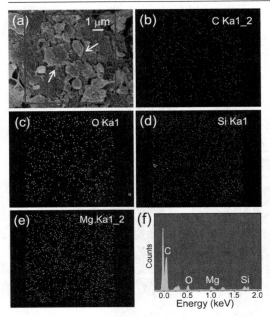

Figure 4. FE-SEM micrographs of the collagen fibers (a) before and (b) after addition of Laponite. (c) AFM images of the collagen fibrils and (d) the image at a higher magnification. The inset shows the AFM image for D-periodic banding patterns of the collagen. White arrows indicate Laponite clay nanoplatelets.

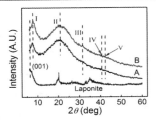

Figure 6. WAXD patterns spectra of Laponite clay powder and the collagen fibers (A) before and (B) after introduction of Laponite.

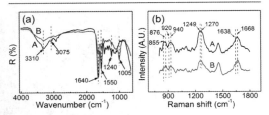

Figure 7. (a) ATR-FTIR and (b) Raman spectra of collagens (A) before and (B) after addition of Laponite.

Figure 5. (a) FE-SEM image of the collagen fibers after introduction of Laponite. White arrows show the depositions of clay nanoplatelets onto the collagen fibers. EDX elemental mappings of (b) carbon (C, red), (c) oxygen (O, yellow), (d) silicon (Si, green), and (e) magnesium (Mg, blue) are present simultaneously within the collagen fibers. (f) The corresponding EDX spectrum of the collagen fibers.

can be observed, which deposit on or between the collagen fiber strands (Figure 4b). Figure 4c,d shows that the assembled collagen fibrils maintain their typical native D-periodic banding patterns (∼65 nm) with the clay nanoplatelets dispersed between the fibrils. It is accepted that, only when the polypeptide chains of collagen molecules are packed together in an orderly way to maintain their helical conformation, the D-periodic structure can be formed.[21] Therefore, we can conclude that the primary and conformational structures of collagen molecules are not destructed by the introduction of Laponite.

The distribution of Laponite on the surface of the wet-white leather matrix (pink area in Figure 5a) has further been investigated by FE-SEM/EDX. It should be noted that the small nanoplatelets cannot be clearly emerged in the FE-SEM

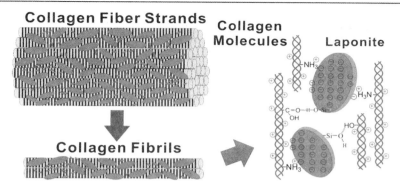

Figure 8. Schematic representation showing the binding of Laponite clay nanoparticles in type I collagen and noncovalent interactions between collagen and Laponite.

images under the finite magnification (Figure 5a). And the observed Laponite aggregates show a little larger than that inside the collagen fibers (Figure 4b). It is reasonable because they are bound onto the surface of the leather matrix. Carbon (Figure 5b) and oxygen (Figure 5c) in the EDX mappings are mainly derived from the organic components of collagen fibers. Silicon (Figure 5d) and magnesium (Figure 5e), two feature elements of Laponite, are present simultaneously within the collagen fibers (Figure 5f). The silicon and magnesium mappings also suggest a relatively uniform distribution of Laponite within the collagen fibers.

The WAXD analysis results are shown in Figure 6. From the position of the maxima in reciprocal space, the d values are obtained by Bragg's equation. It shows that Laponite-treated collagen fiber shows no diffraction peak at $d \sim 1.60$ nm (001) assigned to the interlayer spacing of clay nanoplatelets,[31] indicating good nanodispersion of Laponite within the collagen fibers.[14,32] Moreover, the five characteristic diffraction peaks are observed in the spectra of pure collagen fibers and the leather matrix, namely, the sharp peak I at 1.18 nm assigned to the intermolecular lateral packing within collagen fibrils,[33] the broad peak II at around 0.426 nm related to the amorphous scatter of the unordered components of collagen fibers, peak III at ~0.280 nm corresponding to the axial rise distance between amino acid residues along collagen triple helices or helical rise per residue,[34] and peak IV (0.241 nm) and peak V (0.20 nm) attributed to the axial translation values for amino acid residues in N and C telopeptide, respectively.[35] The apparently unchanged peak I indicates that Laponite nanoplatelets do not enter the space between collagen molecules within microfibrils, thus maintaining the D-periodic banding patterns as discussed above. Obviously, the interactions of Laponite with collagen are different from traditional Cr^{3+} salts, which can form complex bonding with side-chain carboxyls of aspartic and glutamic acids, leading to the distortions of triple-helical conformation and intermolecular lateral packing of the collagen.[21] And, therefore, the WAXD diffraction patterns changed accordingly.[21] As far as Laponite concerned, the reactive center of the collagen that can be reached is the surface of collagen fibrils, and the interactions cannot exclude hydrogen bonds, in addition to ionic bonding. As such, solo Laponite tanning gives inadequate leathering effect, although it can enhance the mechanical properties of the hide collagen as proved in our previous study.[18]

It is known that the intact fibrillar type I collagen has a special supercoiled triple-helical conformation,[36] and its backbone structure with high proportion of amide and imide groups can be well characterized by ATR-FTIR[37] and Raman spectroscopy.[38] Figure 7a depicts FTIR spectra of the collagen before and after the introduction of Laponite. The typical peaks of 3310 and 3075 cm^{-1} (N–H stretching of amides A and B), 1640 cm^{-1} (amide I, peptide C=O stretching), 1550 cm^{-1} (amide II, N–H-bending vibrations coupled to C–N stretching), and 1240 cm^{-1} (amide III, C–N stretching, N–H bending, and –CH$_2$ wagging) indicate the backbone structure of collagen. It is found that the positions of these typical peaks do not change and the amide I peak directly relates to the triple-helical structure of collagen maintains at the same position after the introduction of Laponite. A new and strong peak at 1005 cm^{-1} ascribed to the asymmetric-stretching vibration of Si–O–Si in Laponite[39] is observed in the collagen fibers. Moreover, the IR peak absorption ratio of amide III (1240 cm^{-1}) to 1450 cm^{-1}, a measure of the structure integrity of the collagen triple helix,[40] is close to 1.0, and the differences between wavenumbers of amides I and II reflecting the presence of denaturation process of α-helix chains in collagen molecules are all below 100 cm^{-1}.[41] As given in the Raman spectra (Figure 7b), the peaks at 1668 and 1638 cm^{-1} (the amide I peak, the peptide carbonyl-stretching vibration), 1270 and 1249 cm^{-1} (the amide III peak, N–H in plane deformation coupled to C–N-stretching mode), and 940 cm^{-1} (C–C stretching, ascribed to the peptide backbone) support the presence of the helical conformation in the collagen molecules.[42] The peaks at 920 and 855 cm^{-1} corresponding to proline (Pro) and 876 cm^{-1} to hydroxyproline (Hyp) are observed for strong Raman scattering of saturated side-chain rings. Note that there are no obvious spectral changes in these typical peaks, indicating that the presence of Laponite does not destroy the triple-helical structure of collagen. The ATR-FTIR and Raman spectra results further reveal the noncovalent interactions between Laponite and the collagen.

Combining these microstructural analyses and the influence of Laponite in the increasing the thermal stability and mechanical properties of collagen fibers,[18] the interactions between collagen and Laponite nanoparticles can be combined ionic and hydrogen bonding, as graphically illustrated in Figure 8. The former is originated from the positively charged amino groups of the collagen in the initial tanning float at pH ~ 3.0

(<IEP ~ 6.8), which give rise to ionic bonding with negatively charged Laponite. Whereas for the latter, because Laponite also contains Si—OH groups, it enables them to form hydrogen bonds with collagen.

■ CONCLUSIONS

In the present study, we have studied the diffusion behaviors of Laponite clay nanoparticles and its potential binding centers in the hierarchical structure of the collagen fiber network. The results show that Laponite exhibits high affinity for the collagen and the adsorption behaviors fit well with the Langmuir model. The introduction of Laponite into the collagen fibers can lead to a shift in the IEP from ~6.8 to ~4.5, indicating that the driving force for the penetration and binding of Laponite must involve electrostatic interaction. Owing to the small size of the nanoplatelets and its favorable nanodispersibility, Laponite can be well distributed and bound onto the collagen fibrils without destructing their native D-periodic banding patterns. Moreover, the triple-helical conformation of collagen molecule is also well maintained. These results reveal that the interactions between collagen and Laponite can be the combination of electrostatic interaction and hydrogen bonding.

■ AUTHOR INFORMATION

Corresponding Authors
*E-mail: tongai@cuhk.edu.hk (T.N.).
*E-mail: wlin@scu.edu.cn (W.L.).

ORCID
To Ngai: 0000-0002-7207-6878

Notes
The authors declare no competing financial interest.

■ ACKNOWLEDGMENTS

Financial support from National Natural Science Foundation (NNSF) of China (21476148), Innovation Team Program of Science & Technology Department of Sichuan Province (Grant no. 2017TD0010), National Key R&D Program of China (2017YFB0308402) and Scientific Research Program of Shaanxi Provincial Education Department (Program no.18JK0611) are gratefully acknowledged. T.N. acknowledges support by the National Natural Science Foundation (NNSF) of China (21574110).

■ REFERENCES

(1) Engel, J.; Bächinger, H. P. Structure, Stability and Folding of the Collagen Triple Helix. *Top. Curr. Chem.* **2005**, *247*, 7−33.
(2) Shoulders, M. D.; Raines, R. T. Collagen Structure and Stability. *Annu. Rev. Biochem.* **2009**, *78*, 929−958.
(3) Reyna-Valencia, A.; Chevallier, P.; Mantovani, D. Development of a Collagen/Clay Nanocomposite Biomaterial. *Mater. Sci. Forum* **2012**, *706−709*, 461−466.
(4) Reich, G. What is Leather? The Structure and Reactivity of Collagen. *From Collagen to Leather − The Theoretical Background*; BASF Service Center Media and Communications: Ludwigshafen, Germany, 2007; pp 11−16.
(5) Covington, A. D. Modern Tanning Chemistry. *Chem. Soc. Rev.* **1997**, *26*, 111−126.
(6) Brown, E. M.; Shelly, D. C. Molecular Modeling Approach to Vegetable Tanning: Preliminary Results for Gallotannin Interactions with the Collagen Microfibril. *J. Am. Leather Chem. Assoc.* **2011**, *106*, 145−152.
(7) Guo, J.; Huang, X.; Wu, C.; Liao, X.; Shi, B. The Further Investigation of Tanning Mechanisms of Typical Tannages by Ultraviolet-Visible and Near Infrared Diffused Reflectance Spectrophotometry. *J. Am. Leather Chem. Assoc.* **2011**, *106*, 226−231.
(8) Zhang, Y.; Ingham, B.; Leveneur, J.; Cheong, S.; Yao, Y.; Clarke, D. J.; Holmes, G.; Kennedy, J.; Prabakar, S. Can Sodium Silicates Affect Collagen Structure During Tanning? Insights from Small Angle X-ray Scattering (SAXS) Studies. *RSC Adv.* **2017**, *7*, 11665−11671.
(9) Liu, Y.; Chen, Y.; Yao, J.; Fan, H.; Shi, B.; Peng, B. An Environmentally Friendly Leather-Making Process Based on Silica Chemistry. *J. Am. Leather Chem. Assoc.* **2010**, *105*, 84−93.
(10) Brunier, B.; Sheibat-Othman, N.; Chniguir, M.; Chevalier, Y.; Bourgeat-Lami, E. Investigation of Four Different Laponite Clays as Stabilizers in Pickering Emulsion Polymerization. *Langmuir* **2016**, *32*, 6046−6057.
(11) Elzbieciak, M.; Wodka, D.; Zapotoczny, S.; Nowak, P.; Warszynski, P. Characteristics of Model Polyelectrolyte Multilayer Films Containing Laponite Clay Nanoparticles. *Langmuir* **2010**, *26*, 277−283.
(12) Brunier, B.; Sheibat-Othman, N.; Chevalier, Y.; Bourgeat-Lami, E. Partitioning of Laponite Clay Platelets in Pickering Emulsion Polymerization. *Langmuir* **2016**, *32*, 112−124.
(13) Li, C.; Mu, C.; Lin, W.; Ngai, T. Gelatin Effects on the Physicochemical and Hemocompatible Properties of Gelatin/PAAm/Laponite Nanocomposite Hydrogels. *ACS Appl. Mater. Interfaces* **2015**, *7*, 18732−18741.
(14) Rao, Y. Gelatin-Clay Nanocomposites of Improved Properties. *Polymer* **2007**, *48*, 5369−5375.
(15) Bao, Y.; Ma, J.; Wang, Y. Preparation of Acrylic Resin/Montmorillonite Nanocomposite for Leather Tanning Agent. *J. Am. Leather Chem. Assoc.* **2009**, *104*, 352−358.
(16) Su, D.; Wang, C.; Cai, S.; Mu, C.; Li, D.; Lin, W. Influence of Palygorskite on the Structure and Thermal Stability of Collagen. *Appl. Clay Sci.* **2012**, *62−63*, 41−46.
(17) Shi, J.; Puig, R.; Sang, J.; Lin, W. A Comprehensive Evaluation of Physical and Environmental Performances for Wet-White Leather Manufacture. *J. Cleaner Prod.* **2016**, *139*, 1512−1519.
(18) Shi, J.; Ren, K.; Wang, C.; Wang, J.; Lin, W. A Novel Approach for Wet-White Leather Manufacture Based on Tannic Acid-Laponite Nanoclay Combination Tannage. *J. Soc. Leather Technol. Chem.* **2016**, *100*, 25−30.
(19) He, L.; Mu, C.; Shi, J.; Zhang, Q.; Shi, B.; Lin, W. Modification of Collagen with a Natural Cross-Linker, Procyanidin. *Int. J. Biol. Macromol.* **2011**, *48*, 354−359.
(20) Vidal, C. M. P.; Leme, A. A.; Aguiar, T. R.; Phansalkar, R.; Nam, J.-W.; Bisson, J.; McAlpine, J. B.; Chen, S.-N.; Pauli, G. F.; Bedran-Russo, A. Mimicking the Hierarchical Functions of Dentin Collagen Cross-Links with Plant Derived Phenols and Phenolic Acids. *Langmuir* **2014**, *30*, 14887−14893.
(21) Wu, B.; Mu, C.; Zhang, G.; Lin, W. Effects of Cr^{3+} on the Structure of Collagen Fiber. *Langmuir* **2009**, *25*, 11905−11910.
(22) Metreveli, G.; Kaulisch, E. M.; Frimmel, F. H. Coupling of a Column System with ICP-MS for the Characterisation of Colloid-Mediated Metal (Loid) Transport in Porous Media. *Acta Hydrochim. Hydrobiol.* **2005**, *33*, 337−345.
(23) Zembala, M.; Adamczyk, Z. Measurements of Streaming Potential for Mica Covered by Colloid Particles. *Langmuir* **2000**, *16*, 1593−1601.
(24) Spettmann, D.; Eppmann, S.; Flemming, H. C.; Wingender, J. Visualization of Membrane Cleaning Using Confocal Laser Scanning Microscopy. *Desalination* **2008**, *224*, 195−200.
(25) Tawari, S. L.; Koch, D. L.; Cohen, C. Electrical Double-Layer Effects on the Brownian Diffusivity and Aggregation Rate of Laponite Clay Particles. *J. Colloid Interface Sci.* **2001**, *240*, 54−66.
(26) Reich, G. Physical and Chemical Processes on Collagen and Its Transformation into the Leather Matrix. *From Collagen to Leather − The Theoretical Background*; BASF Service Center Media and Communications: Ludwigshafen, Germany, 2007; pp 42−43.
(27) Sun, J.; Rao, S.; Su, Y.; Xu, R.; Yang, Y. Magnetic Carboxymethyl Chitosan Nanoparticles with Immobilized Metal Ions for Lysozyme Adsorption. *Colloid Surf., A* **2011**, *389*, 97−103.

(28) Tsai, D. H.; DelRio, F. W.; Keene, A. M.; Tyner, K. M.; MacCuspie, R. I.; Cho, T. J.; Zachariah, M. R.; Hackley, V. A. Adsorption and Conformation of Serum Albumin Protein on Gold Nanoparticles Investigated Using Dimensional Measurements and in Situ Spectroscopic Methods. *Langmuir* **2011**, *27*, 2464−2477.

(29) Liao, X.; Zhang, M.; Shi, B. Collagen-Fiber-Immobilized Tannins and Their Adsorption of Au(III). *Ind. Eng. Chem. Res.* **2004**, *43*, 2222−2227.

(30) Mahmoudi, M.; Lynch, I.; Ejtehadi, M. R.; Monopoli, M. P.; Bombelli, F. B.; Laurent, S. Protein-Nanoparticle Interactions: Opportunities and Challenges. *Chem. Rev.* **2011**, *111*, 5610−5637.

(31) Lotsch, B. V.; Ozin, G. A. All-Clay Photonic Crystals. *J. Am. Chem. Soc.* **2008**, *130*, 15252−15253.

(32) Delhom, C. D.; White-Ghoorahoo, L. A.; Pang, S. S. Development and Characterization of Cellulose/Clay Nanocomposites. *Composites, Part B* **2010**, *41*, 475−481.

(33) Sionkowska, A.; Wisniewski, M.; Skopinska, J.; Kennedy, C. J.; Wess, T. J. Molecular Interactions in Collagen and Chitosan Blends. *Biomaterials* **2004**, *25*, 795−801.

(34) Beck, K.; Brodsky, B. Supercoiled Protein Motifs: The Collagen Triple-Helix and the α-Helical Coiled Coil. *J. Struct. Biol.* **1998**, *122*, 17−29.

(35) Orgel, J. P.; Wess, T. J.; Miller, A. The *in Situ* Conformation and Axial Location of the Intermolecular Cross-Linked Non-Helical Telopeptides of Type I Collagen. *Structure* **2000**, *8*, 137−142.

(36) Brodsky, B.; Persikov, A. V. Molecular Structure of the Collagen Triple Helix. *Adv. Protein Chem.* **2005**, *70*, 301−339.

(37) Doyle, B. B.; Bendit, E. G.; Blout, E. R. Infrared Spectroscopy of Collagen and Collagen-like Polypeptides. *Biopolymers* **1975**, *14*, 937−957.

(38) Frushour, B. G.; Koenig, J. L. Raman Scattering of Collagen, Gelatin, and Elastin. *Biopolymers* **1975**, *14*, 379−391.

(39) Pálková, H.; Madejová, J.; Zimowska, M.; Serwicka, E. M. Laponite-Derived Porous Clay Heterostructures: II. FTIR Study of the Structure Evolution. *Microporous Mesoporous Mater.* **2010**, *127*, 237−244.

(40) Andrews, M. E.; Murali, J.; Muralidharan, C.; Madhulata, W.; Jayakumar, R. Interaction of Collagen with Corilagin. *Colloid Polym. Sci.* **2003**, *281*, 766−770.

(41) Albu, M. G.; Ghica, M. V.; Leca, M.; Popa, L.; Borlescu, C.; Cremenescu, E.; Giurginca, M.; Trandafir, V. Doxycycline Delivery from Collagen Matrices Crosslinked with Tannic Acid. *Mol. Cryst. Liq. Cryst.* **2010**, *523*, 97−105.

(42) Ikoma, T.; Kobayashi, H.; Tanaka, J.; Walsh, D.; Mann, S. Physical Properties of Type I Collagen Extracted from Fish Scales of Pagrus Major and Oreochromis Niloticas. *Int. J. Biol. Macromol.* **2003**, *32*, 199−204.

Novel Wet-White Tanning Approach Based on Laponite Clay Nanoparticles for Reduced Formaldehyde Release and Improved Physical Performances

Jiabo Shi,[†] Chunhua Wang,[‡] Liyuan Hu,[‡] Yuanhang Xiao,[‡] and Wei Lin*[,†,‡]

[†]College of Bioresources Chemical and Materials Engineering and National Demonstration Center for Experimental Light Chemistry Engineering Education, Shaanxi University of Science & Technology, No. 6 Xuefu Zhonglu, Weiyang District, Xi'an 710021, China

[‡]National Engineering Laboratory for Clean Technology of Leather Processing, Sichuan University, No. 24 South Section 1, Yihuan Road, Chengdu 610065, China

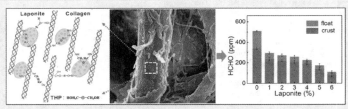

ABSTRACT: In this work, a novel eco-friendly wet-white tanning approach based on tetrakis(hydroxymethyl) phosphonium sulfate (THPS) and synthetic Laponite clay nanoparticles has been developed to reduce potential HCHO risk as well as to improve leather performances. Our results indicate that the hydrothermal stability enhancement of the leather and the HCHO release are closely related to the dosage of THPS and final float pH in the THPS-based wet-white tanning process. The introduction of Laponite can cause distinct increase in shrinkage temperature (T_s) of the combination tanned leather and reduced HCHO contents, implying the presence of synergistic effects between THPS and Laponite. The wet-white tanning system of 2.5% THPS combined 3% Laponite by two-bath method at final pH ∼4.5 is thus optimized, conferring the leather with a T_s above 85 °C. Laponite clay nanoplatelets can be evenly bound between collagen fibrils without altering the native D-periodic banding patterns. Moreover, the novel combination tanning not only improves yellowing resistance and lightfastness but also enhances strength properties of the wet-white leather. These findings provide a potential application of Laponite to meet growing demands for reasonable tanning materials and related technologies toward eco-friendly leather manufacture.

KEYWORDS: *Wet-white tanning, THPS, Laponite nanoclay, HCHO release, Eco-friendly leather*

■ INTRODUCTION

In recent decades, wet-white tanning systems based on nonmetal materials, such as aldehyde compounds,[1,2] polymeric aldehydes,[3] phosphonium salts,[4] and oxazolidine compounds[5] have been extensively explored as eco-friendly and sustainable alternatives to conventional chrome tanning because of increasing issues with chromium-containing leather wastes and tannery sludges.[6,7] The application of these alternatives can also avoid potential Cr(VI) formation in leather processing.[8] However, formaldehyde (HCHO) can be continuously released during post-tanning processes and storage of leather articles as HCHO was generally employed as the raw material for the production of these alternatives.[9] It is known that HCHO has been classified as a potential human carcinogen by the International Agency for Research on Cancer (IARC).[10] In 2009, the European Union has promulgated the Eco-label (2009/563/EC) for footwear in which the amount of HCHO should not exceed the limit of 150 ppm. Stricter limitation requirements for HCHO in leather articles have been proposed as 75 ppm in the Law 112 of Japan and the Eco-leather label recommended by the China Leather Industry Association.[11] Recently, the residual HCHO in textile, leather, and other consumer products is facing more stringent environmental regulations. Therefore, it is imperative to develop reasonable tanning materials and related tanning technologies without potential risks of HCHO to human health toward eco-friendly leather manufacturing.

Tetrakis(hydroxymethyl) phosphonium sulfate (THPS) is an organophosphine derivative as a typical water-soluble quaternary phosphonium salt. It has been utilized in a wide variety of industries including wastewater treatment agent,[12] and environmentally friendly biocide due to its low environ-

Received: September 21, 2018
Revised: November 10, 2018
Published: November 16, 2018

mental toxicity and no potential bioaccumulation.[13] In the leather industry, it is highly considered as an effective nonmetal tanning agent for wet-white leather manufacture, and the aldehydic reaction mechanism between active hydroxymethyl groups of THP molecules and functional amino groups of collagen is generally accepted.[14,15] Moreover, it can confer leathers with fine hydrothermal stability and promote the uptake of anionic fatliquoring agents and dyestuffs in post-tanning procedures. Nevertheless, THPS tanning has several shortages such as potential HCHO release,[16] poor yellowing resistance, and lightfastness of the leathers. The combination tannages using THPS with vegetable tannins,[17] aluminum salt,[18] and other chrome-free tanning materials have attracted considerable interest to acquire desirable physical performances and preferable ecological benefits.

Laponite nanoclay is a uniform synthetic disk-shaped hectorite clay with a thickness of 1 nm, a diameter of 25–30 nm, and permanent negative surface charges.[19] It has been commercially used in various waterborne products attributed to its excellent colloidal properties, nontoxicity, and no potential risks to human health.[20] It is reported that the clay nanoparticles can be well-exfoliated and dispersed into the gelatin–Laponite nanocomposite films exhibiting enhanced physical performances.[21] Our recent studies reveal that Laponite can be explored as a promising combination tanning agent for chrome-free leather making[22] because of its good penetration into collagen fibers and strong ionic binding with collagen.[23] In addition, natural nanoclay such as montmorillonite[24] and attapulgite[25] can be surface-functionalized by quaternary phosphonium salts to obtain long-term antibacterial activity, showing the interactions between them. Meanwhile, nanoclays possess high specific surface area, excellent nanoporosity, and plentiful surface hydroxyl groups which may provide sufficient reaction sites for the adsorption of HCHO.[26]

Herein, we introduce Laponite nanoclay into THPS-based wet-white tanning for the production of eco-friendly leather. Synergistic interactions between THPS and Laponite and the influence of Laponite on the HCHO release after tanning have been mainly studied. The microstructure and physical performances of the leather matrix, especially lightfastness and strength properties, have also been examined. Our aim is to develop a feasible wet-white tanning approach by integrating the advantages of THPS and the nanoclay toward eco-friendly leather manufacturing.

■ EXPERIMENTAL SECTION

Materials. Pickled goatskin pelts were used as raw materials for leather processing. The dosages of chemicals were all based on the weight of pelts. THPS solution (75%, w/w) was purchased from Xiya Chemical Industry Co., Ltd. (China). Laponite RDS, a synthetic disc-shaped nanosilicate consisting of clay nanoparticles incorporating an organic pyrophosphate peptiser ($Na_4P_2O_7$), was provided from BYK Additives & Instruments Ltd. (U.K.). Its chemical composition (dry basis) is as follows: SiO_2 (54.5 wt %), MgO (26.0 wt %), Li_2O (0.8 wt %), Na_2O (5.6 wt %), and P_2O_5 (4.4 wt %), offered by the supplier and often cited in research articles.[19] The chemicals used for the leather processing were of commercial grade, and the others were of analytical ones purchased from Chengdu Kelong Chemical Co., Ltd.

Wet-White Tanning Process. The solo THPS tanning process was based on a previously reported method.[17] Specifically, the pickled pelts were first immersed in 100% pickle liquors (8% NaCl and 1.25% H_2SO_4) at 25 °C and repickled by HCOOH solution (1.0%, w/w) to adjust the pH of pelts to ~3.0 for 1 h. Then the repickled pelts were tanned with 0.5–3.5% THPS, respectively, and the drum was run for 2 h to give a complete penetration of THPS. Final pH of the floats and the crust leather was adjusted to 3.0–5.5 with an interval of 0.5 by $NaHCO_3$ solution (1.5%, w/w). The obtained crust leathers were then washed with 200% water and piled for 24 h. The hydrothermal stability of the leathers was tested, and the HCHO contents in the crusts and the tanned floats were determined, respectively.

As for the combination tanning for wet-white leather, the picked pelts were initially tanned with 2.5% THPS according to the above optimized solo tanning approach. The obtained crust leathers were then rinsed, drained, and respectively tanned with 1–6% Laponite, namely, two-bath method. Also, successive THPS and Laponite tanning in one bath was also studied for comparison. After the bath was run for another 2 h, the floats were finally adjusted to pH ~4.5. The resulting leathers were then washed and piled for 24 h. Post-tanning processing was conducted in accordance with normal procedures including retanning and fatliquoring.[16] The leather properties and the HCHO contents were subsequently determined. The solo Laponite tanned leather[27] was used as control.

Measurement of Shrinkage Temperature. Shrinkage temperature (T_s), a measure of hydrothermal stability of the leather, was determined by a shrinkage tester according to the official method.[22] The 10 mm × 60 mm specimen from leather samples was held in water and heated at a rate of 2 °C/min. The point at which a specimen shrinks was recorded as T_s. Each experimental plot was obtained from an average of three samples.

Determination of HCHO Contents. Concentrations of HCHO in crusts and floats from the tanning process were determined according to the official approach recommended by the International Union of Leather Technologists and Chemists Societies (IULTCS): IUC 19 (in accordance with ISO 17226-1:2008) using high-performance liquid chromatography (HPLC).[28] A 2.0 g leather sample was cut into pieces (5 mm × 5 mm) and extracted at 40 °C with 50 mL of 0.1% sodium dodecyl benzenesulfonate aqueous solution for 60 min. A 5.0 mL extracted solution or 1.0 mL float was mixed and reacted with 2.0 mL of 2,4-dinitrophenyl hydrazine solution at 60 °C, incubating for 30 min. The mixture was cooled to room temperature after derivative reactions. The derivatized solution was filtered through 0.45 μm filters to remove fine particles before the HPLC analysis, performed on an Agilent LC1200 HPLC instrument with an Aglient Eclipse plus C18 column (4.6 mm × 250 mm × 5 μm). The peak area was used for quantitative calculations of HCHO concentration. Each experimental plot was obtained from an average of three samples.

Wide-Angle X-ray Diffraction (WAXD) Studies. Wide-angle X-ray diffraction measurements were performed on a Philips Analytical X'pert ProMPD X-ray diffractometer coupled with a plumbaginous-monochromated Cu Kα irradiation of 0.154 nm. The specimens were scanned at a diffraction angle (2θ) ranging from 5° to 45° at ambient temperature and humidity. The real lattice space d, which represents characteristic structure dimension, can be calculated by $d = \lambda/(2 \sin \theta)$, where λ is the X-ray wavelength and θ is half of the diffraction angle.

Field-Emission Scanning Electron Microscopy (FE-SEM)/Energy-Dispersive X-ray (EDX) Spectroscopy Analysis. Microstructures of leather samples after being lyophilized at −43 °C for 24 h were observed on a FEI Quanta 400 FEG field-emission scanning electron microscope after being sputter-coated with Au. The relative elemental compositions of the corresponding samples were also analyzed by a coupled energy-dispersive X-ray spectroscopy detector.

Atomic Force Microscopy (AFM) Observation. The frozen leather samples with a thickness of 50 μm were dried at room temperature for 24 h in a desiccator with silica gel. Topographic images were acquired on a Shimadzu SPM-9600 atomic force microscope under tapping mode.

Inductively Coupled Plasma-Atomic Emission Spectrometry (ICP-AES) Measurements. The content of Laponite in the wet-white leathers were monitored by ICP-AES by measuring the Mg trace as the characteristic element of Laponite. The measurement of Mg^{2+} concentration was conducted on a PerkinElmer Optima 2100DV spectrometer according to our reported approach.[22] The content of Laponite in leathers was calculated as its commercial composition and expressed as mg/g. Each experimental plot was obtained from an average of three samples.

UV Irradiation Experiments. UV spectra of the wet-white leathers were recorded on a Shimadzu UV-3600 spectrophotometer after UV irradiation for 24 h according to the reported approach.[29] Color parameters of CIE 1976 L^* a^* b^* values and color tristimulus values (XYZ) were analyzed with the coupled color analysis software. Yellowing index (YI) was calculated by YI = 100 × (1.28X − 1.06Z)/Y, where X, Y, and Z values represent color tristimulus values. Each experimental plot was obtained from an average of three samples.

Physical Strength Properties Examination. Physical strength properties of the wet-white leathers were examined using the standard IULTCS methods. Specifically, tear strength, tensile strength, and elongation at break of the leathers were tested using IUP 8 (in accordance with ISO 3377-2:2016) and IUP 6 (in accordance with ISO 3376:2011), respectively. Each experimental plot was obtained from an average of three samples.

RESULTS AND DISCUSSION

Optimization of THPS Tanning System. Figure 1a shows that the T_s of THPS-tanned crust leathers increase

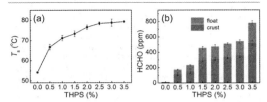

Figure 1. (a) T_s of crust leathers as a function of THPS dosage in solo THPS tanning at final pH 4.5; (b) corresponding concentrations of HCHO in both crusts and tanned floats.

markedly with increasing the dosage of THPS. This can be mainly attributed to covalent cross-linking formation between active hydroxymethyl groups of THP molecules and functional amino groups of collagen side chains.[14] At the dosage of 2.5% based on the weight of pickled pelts, T_s of ∼79 °C is achieved. Whereas further increasing of THPS does not result in an obvious increase in the T_s, the concentrations of HCHO in both the crust leathers and floats increase distinctly (Figure 1b). It suggests the release of HCHO from unbound THPS. This is understandable because HCHO is used as the raw material for the production of THPS. Because of its structural instability and aldehydic reaction with hide collagen, it can be released from unbound THPS and leather matrix.[14] Thus, 2.5% THPS is selected as the dosage in the subsequent combination tanning experiments.

Figure 2a,b shows the effect of final pH on the T_s of the leathers tanned with 2.5% THPS and the concentrations of HCHO, respectively. It is obvious that T_s increases with the pH from 3.0 to 5.5 (Figure 2a). This is because collagen is an polyampholyte, and thus its ionization and reactivity are greatly influenced by the pH of floats, and high final pH can facilitate the fixation of aldehydic tanning materials with collagens.[14] Note that the HCHO concentrations in both crusts and floats also apparently increase at pH 5.0−5.5 (Figure 2b), which may be due to the fact that higher pH is favorable for the dissociations of THPS to release HCHO.[30] Therefore, in consideration of the tanning effects and potential HCHO release, tanning with 2.5% THPS at final pH ∼4.5 is preferred in the following combination tanning experiments.

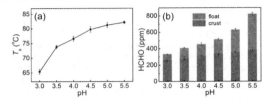

Figure 2. (a) Final pH dependence of T_s of crust leathers and (b) corresponding HCHO concentrations in both crusts and floats by 2.5% THPS tanning.

Optimization of Wet-White Combination Tanning. Figure 3a presents the T_s of crust leathers tanned by 2.5%

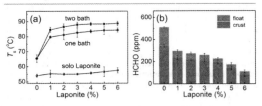

Figure 3. (a) T_s of wet-white crust leathers produced by 2.5% THPS−Laponite combination tanning system. (b) Concentrations of HCHO in the crusts and tanning floats from the combination tanning process.

THPS combined with Laponite nanoclay. It is clear that solo Laponite tanning gives a limited increase in the T_s (less than 5 °C) because of a minor contribution of ionic bonding between the negatively charged clay nanoplatelets and the positively charged amino groups of collagens.[23] By contrast, THPS−Laponite combination tanning confers the crusts a significant increase in T_s, much higher than solo THPS (∼79 °C) or Laponite tanning (∼55 °C), implying the presence of synergistic interactions of THPS−Laponite. Besides, the two-bath method, that is, first 2.5% THPS tanning, then draining, and 3% Laponite tanning, shows ∼5 °C higher in T_s than the one-bath method (no draining). It means that, in the latter case, the addition of Laponite would preferentially bind with the THP in the float to form amorphous precipitate, thus leading to insufficient penetration and bindings of the clays into collagen fibers. Additionally, further increasing the dosage of Laponite does not give a distinct enhancement in T_s of the leathers.

Figure 3b shows that with the increasing amount of Laponite, the concentrations of HCHO in both crusts and tanning floats decrease distinctly. The reason can be related to the synergistic effects of THPS−Laponite in the formation of leather matrix, on the one hand, and on the other hand, to the nanoporous structure of Laponite with strong adsorption capacity and good affinity for HCHO molecules.[31] Consequently, wet-white combination tanning approach with 2.5% THPS and 3% Laponite in separate floats (two-bath method) is optimized.

WAXD Analysis. The dispersion of Laponite nanoclay into the collagen fibers of THPS tanned wet-white leathers is studied. The WAXD patterns are given in Figure 4. Note that the diffraction peak at d ∼ 1.60 nm (001) is assigned to the interlayer spacing of the clay nanoplatelets.[32] The peak is absent in the pattern of the combination tanned wet-white

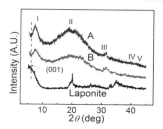

Figure 4. WAXD patterns of Laponite nanoclay powder and wet-white leathers tanned by (A) THPS and (B) THPS−Laponite.

leather, indicating good exfoliation and nanodispersion of the clay into the collagen fibers of the leather matrix.[33] Moreover, five characteristic diffraction peaks of the collagen fibrils are observed in the pattern of the combination tanned leather matrix, that is, the sharp peak I at 1.18 nm corresponding to the intermolecular lateral packing within the collagen fibrils,[34] the broad peak II at ∼0.42 nm ascribed to the amorphous scatter of the unordered components of collagen fibers, peak III at ∼0.28 nm attributed to the axial rise distance between amino acid residues along collagen triple helices or helical rise per residue,[35] and peak IV (0.24 nm) and peak V (0.20 nm) related to the axial translation values for amino acid residues in N and C telopeptide, respectively.[36] The unaffected peak I in the wet-white leather matrix suggests that the introduced clay nanoparticles cannot enter into collagen microfibrils in the tanning process, which is in accordance with our previous report.[23]

SEM and AFM Observations. We have combined the results from FE-SEM and AFM in Figure 5 to study the influence of Laponite nanoclay on the microstructure of collagens in the wet-white leather matrix. Figure 5a shows thick and isolated collagen fiber bundles in the leather matrix because of the dehydration effects of the nanosilicates on the collagens. Further amplified FE-SEM image reveals the aggregates of clay nanoplatelets with diameters of 100−300 nm deposited on or between the collagen fiber strands (Figure 5b). Moreover, AFM image (Figure 5c) indicates that the assembled collagen fibrils can maintain native D-periodic banding patterns (∼65 nm) with the clay nanoplatelets well-dispersed between or on the collagen fibrils. It is reported that only when the polypeptide chains of collagen molecules are packed together in an orderly way to maintain their helical conformation, the periodic structure can be formed.[35] It can therefore be concluded that the fibrillar structure of collagens is not affected by adding Laponite into the collagen fibers of the leather matrix, which further agrees with the WAXD results. According to the enhancement of the hydrothermal stability of collagen fibers and the microstructural morphologies, the synergistic interactions of THPS−Laponite in the combination tanning can be attributed to multi-cross-linking effects as graphically illustrated in Figure 5d. The clay nanoplatelets can not only bind onto the surface of collagen fibrils and cross-link with the collagen molecules through noncovalent interactions, such as hydrogen and ionic bonding, but also create new ionic bonding with the positive tetrakis(hydroxymethyl) phosphonium (THP) as a result of the negative surface charges of the clay nanoplatelets.

ICP-AES and SEM/EDX Analysis. The distribution of Laponite nanoclay into the collagen fibers of wet-white leather has been investigated. Figure 6 shows that the contents of

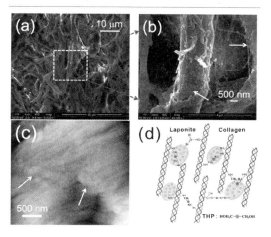

Figure 5. (a) FE-SEM images of wet-white leather tanned by THPS−Laponite at 5000 and (b) 40000 magnifications; (c) AFM image of the corresponding collagen fibrils; (d) schematic representation showing the presence of synergistic interactions between THPS and Laponite with collagen in the combination tanning.

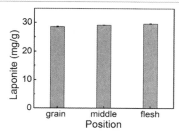

Figure 6. Content of Laponite nanoclay in wet-white leather tanned by 2.5% THPS−3% Laponite as detected by ICP-AES technique.

Laponite in grain, middle, and flesh layers of the leather matrix are all ∼26.0 mg/g, implying homogeneous distribution and favorable binding of Laponite onto the collagen fibers of leather matrix. This can be primarily due to the fact that Laponite clay has small particle size, uniform dimensions, and good nanodispersibility,[19] which make it readily diffuse from the outer floats into collagen fibers. Figure 7a depicts the SEM image of the grain layer of leather matrix in which the hair pores are clearly visible without surface depositions of the clay. And the leather matrix exhibits similar morphologies with an interwoven structure in the middle (Figure 7b) and flesh layers (Figure 7c). Moreover, the corresponding EDX spectra reveal that two feature elements of Laponite, silicon (Si) and magnesium (Mg), are simultaneously detected with similar contents, further confirming homogeneous distribution of Laponite clay within the leather matrix. This is desirable to acquire reasonable and good physical performances in leather tanning.

Physical Performances of Wet-White Leathers. It is known that the yellowing of the leather is primarily associated

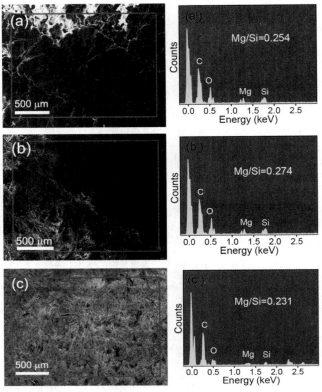

Figure 7. SEM images and the corresponding EDX spectra of 2.5% THPS–3% Laponite combination tanned wet-white leather (a, a′: grain layer; b, b′: middle layer; c, c′: flesh layer).

with the UV irradiation which can cause the changes in collagen primary structure, conformation, microstructure, and physical performances.[37] Parts (a) and (b) of Figure 8 show

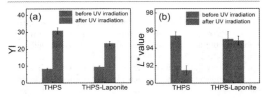

Figure 8. (a) YI and (b) L^* value of wet-white leathers tanned by 2.5% THPS and 2.5% THPS–3% Laponite before and after UV irradiation for 24 h.

the yellowing index (YI) and L^* values of the wet-white leathers tanned by 2.5% THPS and 2.5% THPS-3% Laponite before and after UV irradiation for 24 h, respectively. Obviously, after the irradiation, YI and L^* values of the leather combination tanned by THPS–Laponite change less than those tanned by solo THPS, indicating improved yellowing resistance and lightfastness of the wet-white leather. The increased resistance of the leather to UV irradiation can probably be attributed to the scattering of the incident light by the clay nanoplatelets.[38] Moreover, as given in Table 1, the

Table 1. Physical Strength Properties of Wet-White Leathers

tanning system	tear strength (N·mm)	tensile strength (MPa)	elongation at break (%)
2.5% THPS	56.4 ± 4.0	17.9 ± 0.8	85.5 ± 4.4
2.5% THPS–3% Laponite	85.3 ± 2.0	21.4 ± 0.6	103.9 ± 3.7

combination tanned wet-white leather exhibits enhanced physical strength properties (i.e., tensile strength, tear strength, and elongation at break) over those of solo THPS tanned one. This is in accordance with the hydrothermal stability enhancement of the leather, which is also resulted from the synergistic tanning effects between THPS and Laponite.

■ **CONCLUSIONS**

In this work, a novel eco-friendly wet-white tanning approach based on the combination of THPS and Laponite nanoclay has been established. The results indicate that the hydrothermal stability enhancement and HCHO release of the leather in THPS-based wet-white tanning process are closely related to the dosage of THPS and final pH. The addition of Laponite can provide an obvious increase in T_s of THPS-tanned crust leather, indicating the formation of synergistic effects between THPS and Laponite. Also, the introduction of Laponite can cause apparent reduction of HCHO concentration in both the crusts and tanning floats. The novel wet-white tanning system is optimized as 2.5% THPS with 3% Laponite by a two-bath method at final pH ~4.5, which confers the leather T_s above 85 °C. Laponite clay nanoparticles can be evenly bound on the collagen fibers of the leather matrix without altering native D-periodic banding patterns of the collagen fibrils. Moreover, the novel combination tanning approach can not only improve the yellowing resistance and lightfastness but also enhance the strength properties of the wet-white leather.

■ AUTHOR INFORMATION

Corresponding Author
*E-mail: wlin@scu.edu.cn.
ORCID
Wei Lin: 0000-0003-3880-5026
Notes
The authors declare no competing financial interest.

■ ACKNOWLEDGMENTS

Financial support from the National Natural Science Foundation (NNSF) of China (21476148), Scientific Research Program of Shaanxi Provincial Education Department (18JK0121), Natural Science Basic Research Plan in Shaanxi Province of China (2018JQ2060), and Open Project Program of National Demonstration Center for Experimental Light Chemistry Engineering Education, Shaanxi University of Science & Technology (2018QGSJ02-11) are gratefully acknowledged.

■ REFERENCES

(1) Igl, G.; Wolf, G.; Breth, M.; Carle, J. New developments in wet white tanning technology. *J. Am. Leather Chem. Assoc.* **2001**, 96 (4), 111−119.
(2) Krishnamoorthy, G.; Sadulla, S.; Sehgal, P. K.; Mandal, A. B. Green chemistry approaches to leather tanning process for making chrome-free leather by unnatural amino acids. *J. Hazard. Mater.* **2012**, 215−216 (10), 173−182.
(3) Ballus, Olga; Comes, E.; Palop, R. Wet white tanning by applying a polymeric aldehyde together with dihydroxydiphenylsulfone. *The 31th International Union of Leather Technologists and Chemists Society Congress*, Valencia, Spain, 2011.
(4) Rao, J.; Nair, B. U.; Fathima, N. N.; Kumar, T. P.; Kumar, D. R. Wet white leather processing: A new combination tanning system. *J. Am. Leather Chem. Assoc.* **2006**, 101 (2), 58−65.
(5) Li, J.; Sun, Q.; Wu, C.; Liao, X.; Shi, B. A novel oxazolidine tanning agent and its use in vegetable combination tanning. *J. Soc. Leather Technol. Chem.* **2011**, 95 (4), 165−170.
(6) Qiang, T.; Gao, X.; Ren, J.; Chen, X.; Wang, X. A chrome-free and chrome-less tanning system based on the hyperbranched polymer. *ACS Sustainable Chem. Eng.* **2016**, 4 (3), 701−707.
(7) Lyu, B.; Chang, R.; Gao, D.; Ma, J. Chromium footprint reduction: Nanocomposites as efficient pretanning agents for cowhide shoe upper leather. *ACS Sustainable Chem. Eng.* **2018**, 6 (4), 5413−5423.
(8) Dixit, S.; Yadav, A.; Dwivedi, P. D.; Das, M. Toxic hazards of leather industry and technologies to combat threat: A review. *J. Cleaner Prod.* **2015**, 87 (1), 39−49.
(9) Wolf, G.; Huffer, S. Formaldehyde in leather - A survey. *J. Am. Leather Chem. Assoc.* **2002**, 97 (11), 456−464.

(10) Kim, K. H.; Jahan, S. A.; Lee, J. T. Exposure to formaldehyde and its potential human health hazards. *J. Environ. Sci. Heal. C* **2011**, 29 (4), 277−299.
(11) Sang, J.; Wang, M.; Yu, L.; Zhang, X.; Lin, W. Current situation of chemical management in Chinese leather industry. *Asian J. Ecotoxicol.* **2015**, 10 (2), 123−130.
(12) Okoro, C. C.; Samuel, O.; Lin, J. The effects of tetrakis-hydroxymethyl phosphonium sulfate (THPS), nitrite and sodium chloride on methanogenesis and corrosion rates by methanogen populations of corroded pipelines. *Corros. Sci.* **2016**, 112, 507−516.
(13) Xu, D.; Li, Y.; Gu, T. A synergistic D-tyrosine and tetrakis hydroxymethyl phosphonium sulfate biocide combination for the mitigation of an SRB biofilm. *World J. Microbiol. Biotechnol.* **2012**, 28 (10), 3067−3074.
(14) Covington, A. D. Other tannages. In *Tanning Chemistry: The Science of Leather*; Royal Society of Chemistry: 2009.
(15) Chung, C.; Lampe, K. J.; Heilshorn, S. C. Tetrakis-(hydroxymethyl) phosphonium chloride as a covalent cross-linking agent for cell encapsulation within protein-based hydrogels. *Biomacromolecules* **2012**, 13 (12), 3912−3916.
(16) Li, Y.; Shao, S.; Shi, K.; Lan, J. Release of free formaldehyde in THP salt tannages. *J. Soc. Leather Technol. Chem.* **2008**, 92 (4), 167−169.
(17) Aravindhan, R.; Madhan, B.; Rama, R. Studies on tara-phosphonium combination tannage: Approach towards a metal free eco-benign tanning system. *J. Am. Leather Chem. Assoc.* **2015**, 110 (3), 80−87.
(18) Ren, L.; Wang, X.; Qiang, T.; Ren, Y.; Xu, J. Phosphonium-aluminum combination tanning for goat garment leather. *J. Am. Leather Chem. Assoc.* **2009**, 104 (6), 218−226.
(19) Brunier, B.; Sheibat-Othman, N.; Chniguir, M.; Chevalier, Y.; Bourgeat-Lami, E. Investigation of four different Laponite clays as stabilizers in Pickering emulsion polymerization. *Langmuir* **2016**, 32 (24), 6046−6057.
(20) Elzbieciak, M.; Wodka, D.; Zapotoczny, S.; Nowak, P.; Warszynski, P. Characteristics of model polyelectrolyte multilayer films containing Laponite clay nanoparticles. *Langmuir* **2010**, 26 (1), 277−283.
(21) Rao, Y. Gelatin-clay nanocomposites of improved properties. *Polymer* **2007**, 48 (18), 5369−5375.
(22) Shi, J.; Puig, R.; Sang, J.; Lin, W. A comprehensive evaluation of physical and environmental performances for wet-white leather manufacture. *J. Cleaner Prod.* **2016**, 139, 1512−1519.
(23) Shi, J.; Wang, C.; Ngai, T.; Lin, W. Diffusion and binding of Laponite clay nanoparticles into collagen fibers for the formation of leather matrix. *Langmuir* **2018**, 34 (25), 7379−7385.
(24) Patel, H. A.; Somani, R. S.; Bajaj, H. C.; Jasra, R. V. Preparation and characterization of phosphonium montmorillonite with enhanced thermal stability. *Appl. Clay Sci.* **2007**, 35 (3), 194−200.
(25) Cai, X.; Zhang, J.; Ouyang, Y.; Ma, D.; Tan, S.; Peng, Y. Bacteria-adsorbed palygorskite stabilizes the quaternary phosphonium salt with specific-targeting capability, long-term antibacterial activity, and lower cytotoxicity. *Langmuir* **2013**, 29 (17), 5279−5285.
(26) Bellat, J. P.; Bezverkhyy, I.; Weber, G.; Royer, S.; Averlant, R.; Giraudon, J. M.; Lamonier, J. F. Capture of formaldehyde by adsorption on nanoporous materials. *J. Hazard. Mater.* **2015**, 300, 711−717.
(27) Shi, J.; Ren, K.; Wang, C.; Wang, J.; Lin, W. A novel approach for wet-white leather manufacture based on tannic acid-Laponite nanoclay combination tannage. *J. Soc. Leather Technol. Chem.* **2016**, 100 (1), 25−30.
(28) Bayramoglu, E. E.; Yorgancioglu, A.; Onem, E. Analysis of release of free formaldehyde originated from THP salt tannages in leather by high performance liquid chromatography: Origanum onites essential oil as free formaldehyde scavenger. *J. Am. Leather Chem. Assoc.* **2013**, 108 (11), 411−419.
(29) Liang, Z.; Shan, Z. In situ spectroscopic study on the effect of UV radiation on stabilized collagen fibre: Role of plant tannin. *J. Soc. Leather Technol. Chem.* **2012**, 96 (5), 210−214.

(30) Hu, T.; Williams, T.; Schmidt, J.; James, B.; Cavalin, R.; Lewing, D. Mill trial of the new bleaching agent-THPS. *Pulp Pap.-Canada* **2009**, *110*, 37−42.

(31) Zhu, H. Y.; Lu, G. Q. Engineering the structures of nanoporous clays with micelles of alkyl polyether surfactants. *Langmuir* **2001**, *17* (3), 588−594.

(32) Lotsch, B. V.; Ozin, G. A. All-Clay photonic crystals. *J. Am. Chem. Soc.* **2008**, *130* (46), 15252−15253.

(33) Delhom, C. D.; White-Ghoorahoo, L. A.; Pang, S. S. Development and characterization of cellulose/clay nanocomposites. *Composites, Part B* **2010**, *41* (6), 475−481.

(34) Sionkowska, A.; Wisniewski, M.; Skopinska, J.; Kennedy, C. J.; Wess, T. J. Molecular interactions in collagen and chitosan blends. *Biomaterials* **2004**, *25* (5), 795−801.

(35) Wu, B.; Mu, C.; Zhang, G.; Lin, W. Effects of Cr^{3+} on the structure of collagen fiber. *Langmuir* **2009**, *25* (19), 11905−11910.

(36) Orgel, J. P.; Wess, T. J.; Miller, A. The *in situ* conformation and axial location of the intermolecular cross-linked non-helical telopeptides of type I collagen. *Structure* **2000**, *8* (2), 137−142.

(37) Rabotyagova, O. S.; Cebe, P.; Kaplan, D. L. Collagen structural hierarchy and susceptibility to degradation by ultraviolet radiation. *Mater. Sci. Eng., C* **2008**, *28* (8), 1420−1429.

(38) Essawy, H. A.; Abd El-Wahab, N. A.; Abd El-Ghaffar, M. A. PVC-laponite nanocomposites: Enhanced resistance to UV radiation. *Polym. Degrad. Stab.* **2008**, *93* (8), 1472−1478.